# Anthropology, Islands, and the Search for Meaning in the Anthropocene

Part ethnography, part memoir, and part critical reflection on the Anthropocene, this book examines the ways that islands form and inform human experiences of the everyday and the extraordinary.

Utilizing carefully considered anthropological perspectives drawn from over a decade of anthropological fieldwork, the author employs islands as a complex set of lenses to examine the ways that we are intimately connected, separated, and divided from ourselves, one another, and the planet. Moving across time, place and disciplinary boundaries, this book traces a narrative route from the remote islands of Micronesia to the subarctic expanses of northern Iceland, all in service of gaining a deeper understanding of the cultural resonance of islands.

This book offers the reader a type of ideological travel guide, one that exchanges restaurant reviews and hotel recommendations for pathways of reflection and new modes of seeing and being in the world. It will be of interest to scholars in the social sciences and humanities, and readers from human geography, cultural studies, sociology, philosophy and American studies.

**Justin Armstrong** is Senior Lecturer in Writing and Anthropology at Wellesley College in Wellesley, Massachusetts. His main research interests are experimental ethnography, ethnographic writing, abandoned places, economic anthropology, and the anthropology of islands. He conducts research in Iceland, the Faroe Islands, Newfoundland, and Micronesia. He is the author of a novel, *Wyomings* (2018), along with a number of scholarly articles and book chapters.

## Ocean and Island Studies

Series editors

May Joseph
*Pratt Institute, USA*
Adam Grydehøj
*South China University of Technology, China*
Philip Hayward
*University of Technology Sydney, Australia*
Sudipta Sen
*University of California—Davis, USA*
Lisa Bloom
*University of California—Berkley, USA*
Pamila Gupta
*University of Witwatersrand, Johannesburg, South Africa*

*Ocean and Island Studies* is an interdisciplinary series concerning the role of oceans and islands in thought, theory, practice, and method, past and present. From remote island outposts to bustling island cities and the islands of our dreams, from the expanses and depths of the open sea to coasts, rivers, deltas, lakes, and polar icescapes, oceans and islands are at the core of much contemporary thinking. The books in the series explore the ways in which people use, envision, and construct marine, aquatic, littoral, island, and archipelagic geographies. It is a platform for Blue thinking from across the arts, humanities, social sciences, and environmental sciences. Meditations upon oceanic lyricism and altered states of 'islandness' find their place alongside research into the practicalities of island and coastal economies, infrastructures, and governance.

The series is open to book proposals from all segments of ocean studies, island studies, and related fields. It offers a mix of Shortform titles (i.e. Routledge Focus; 20,000–50,000 words) and Monographs (60,000–80,000 words), that is open to other text types as well.

**Terra Aqua**
The Amphibious Lifeworlds of Coastal and Maritime South Asia
*Edited by Sudipta Sen and May Joseph*

**Anthropology, Islands, and the Search for Meaning in the Anthropocene**
*Justin Armstrong*

For more information please visit: www.routledge.com/Ocean-and-Island-Studies/book-series/OISTU

# Anthropology, Islands, and the Search for Meaning in the Anthropocene

## Justin Armstrong

LONDON AND NEW YORK

First published 2023
by Routledge
4 Park Square, Milton Park, Abingdon, Oxon OX14 4RN

and by Routledge
605 Third Avenue, New York, NY 10158

*Routledge is an imprint of the Taylor & Francis Group, an informa business*

© 2023 Justin Armstrong

*British Library Cataloguing-in-Publication Data*
A catalogue record for this book is available from the British Library

ISBN: 978-1-032-28590-0 (hbk)
ISBN: 978-1-032-28593-1 (pbk)
ISBN: 978-1-003-29758-1 (ebk)

DOI: 10.4324/9781003297581

Typeset in Times New Roman
by Apex CoVantage, LLC

For Heather

# Contents

*Preface: Palm-of-the-Hand Ethnography
and Other Related Considerations*                    viii
*Acknowledgements*                                    xiii

1   Arriving: An Introduction to Island Anthropology       1

2   Wave Glossary: To and from Yap                        12

3   On Becoming an Ethnographic Ghost in the Faroe Islands   20

4   Newfoundland: A Place Apart                           26

5   Iceland I: New Old Dreamworlds                        33

6   Iceland II: Come-from-Away                            40

7   Phantom Islands: Shorelines without Islands           45

8   Hauntological Islands                                 50

9   A Conclusion by Means of Describing Certain
    Lessons That Islands Have Taught Me                   57

*Afterword: The Benefits of Thinking with
Anthropology and Islands*                                61
*Bibliography*                                           63
*Index*                                                  67

# Preface: Palm-of-the-Hand Ethnography and Other Related Considerations

Library drifting was one of the first ways that I travelled to islands. Alone in the dust of the Chancellor Paterson Library at Lakehead University in Thunder Bay, Ontario, my eyes floated across sun-stained spines, landing on anything that slowed my glances for a moment. I was a high school senior pretending to do research for an English paper, looking for books that hadn't been checked out in over a decade. Wandering the stacks, moving between sunspires, wondering about unknowable distances and abstracted places. Here, each forgotten book became an island in an archipelago of knowing, a discrete moment in time and place. Making landfall on their pages allowed me to begin to connect and reconnect the trailing threads of a world of islands that would later envelope me.

It was on one of these grey February afternoons that I stumbled across Yasunari Kawabata's *Palm-of-the-Hand Stories* (1988), which was perhaps the first book that truly made me want to write. In the interior of this cloth-bound island, Kawabata had built miniature universes in tiny stories, most of them less than a page long. I instantly devoured it, and it has remained resolutely on my bookshelf for the last 25 years.

Recently, I returned to the book, not solely as a work of literature, but also as a mode of thinking about my current practice as an ethnographer. These densely packed, yet sparse stories have helped me to form a new approach to research and writing. My aim here has been to take this model and apply it directly to anthropology to produce palm-of-the-hand ethnographies. These small islands of critical reflection and analysis work to distil the experience of the field anthropologist and translate our *aways* into the *homes* of our readers. Kawabata's stories and my ethnographic fragments value sentiment over densities, assembling particles into reflexive, granular traces that evoke subtly shifting impressions of people, places, and things—the Subjects of anthropological discourse. As a way of knowing, seeing and being, palm-of-the-hand ethnography uses the tools of bricolage (Levi-Strauss 1966) and assemblage (Deleuze and Guattari 1988) to echo the ways that we experience and understand the cultural fabric in which we find ourselves wrapped during these last days of history. These small studies in/of ethnography, when taken together, become the archipelagos of cultural inquiry; they are the textures, tones, and colours that bubble up from particular islanded places and times. They are the accumulation of intersections, replacements, reworkings and revisions, cultural scattershot. From my perspective, this model provides a useful structure for thinking *about* and *with* islands where the line between form and function is resonantly unclear.

As such, this book is a collection of micro-essays made out of layers of reflections and meditations on a world made out of islands, both geophysical and ideological. More concretely, it also emerges from over 15 years of anthropological fieldwork. And while the subjects of these micro-essays have often appeared in other places in my research and writing, their particular unpacking in this book is unique. My hope is that these translations of ethnographic encounters might offer readers a renewed and revised sense of attunement as they consider how bounded places form and inform our contemporary late-stage world. These everyday observations explore the ways that we might, as anthropologists, writers, travellers, and human beings, begin to distil our engagements into concentrated, portable meaning and vibrant moments of awareness.

With this model, my goal is to explore the limits of ethnographic sentiment and minimalism in order to answer (a few of) the questions of how and why anthropologists make meaning in our work. Similarly, I employ this mode of analysis to illustrate the notion that our subjects and collaborators are constantly shifting and regenerating, and accordingly so should the ways that we represent them. I feel that this approach is well-suited to the current state of global culture. Movable ideas must be built into compact frameworks, making them relevant and relatable to readers constantly navigating a series of overlapping and competing *-scapes* (Appadurai 1990) made up of an infinitude of endlessly replicating rhizomatic nodes (Deleuze and Guattari 1988). Perhaps this collection can offer its audience some respite from the densities and fractures of everyday life by shifting attentions towards an ethnographic world of islands. Beyond a simple discussion of the intersection of islands and anthropology, this book represents the potential for a renewed sense of wonder and attunement, along with careful considerations of the human and non-human lives lived at the periphery. Perhaps we might think of this as a form of 'deep listening' (Oliveros 2005) to the edges and outer realms. It is important to problematize the roles of Writer and Anthropologist at this intersection, asking how and why these people, places, and things move from *there* to *here*.

## On the Particular Usefulness of Micro-Essays in the Context of Anthropological Inquiry on the Subject of Islands

In each of the chapters that follow, I have framed my discussions as a series of micro-essays, a form of writing that I believe most accurately reflects the topic at hand, and also forms a contemporary and timely mode of engaging with texts, cultural and otherwise. A blunt argument might claim that these short bursts of writing serve an audience whose attention span has all but evaporated in the face of digital technologies, and perhaps that is true to some extent, but this is not my primary aim. This form of densely packed analysis reflects the islands themselves in that they are bounded, discrete, and self-contained. And taken together, they form an archipelago of associated meaning.

I do not claim to be the inventor of this form. Walter Benjamin's *One Way Street* (2021), Renata Adler's *Speedboat* (1976), and Roni Horn's *Island Zombie* (2020) have all provided inspiration and insight into developing a narrative out of vignettes. Most significant, and coming directly from anthropological discourse, are Kathleen Stewart's *Ordinary Affects* (2007) and her recent collaboration with Lauren Berlant, *The Hundreds* (2019), two works that set the stage for what might be called aphoristic anthropology. Similarly, sociologist Allen Shelton's work in *Dreamworlds of Alabama* (2007), along with Michael Taussig's *Palma Africana* (2018) have served as useful models as I consider the form and connectivity of these ethnographic archipelagos.

## Recycling Definitions

Defining "anthropology" and "islands" is not an easy task. These ideas are more slippery than they first appear; they evade and mutate, vanishing under waves of water, politics, culture, and ontology. But allow me to attempt a capture, at least in the context of what follows.

I often tell my students that anthropology is the study of human culture, and that for as many anthropologists as there are in the world, there are an equal number of unique definitions of culture. From where I stand, anthropology is pure translation. It is the forming of the experiences and impressions drawn from fieldwork (that sacred proving ground for our ideas) into relevant and relatable texts, ideally made approachable and available to a wide audience. Here, my intent is to distil and refine my ethnographic encounters with people, places, and things into something graspable and portable. It is the world run through my focused viewpoint with the goal of bringing *away* to *home*. And while I hope that my perspective is valuable and useful to the reader, I also acknowledge that my view is only one of many of equal value and usefulness. And for me, this presence of multiple voices is the essential beauty of anthropology. Just as with any form of inquiry, my anthropological attentions to the world require a subject, and here it is islands, in all of their forms.

And here too we find difficulty in defining "islands" as a singular and concrete entity. As I discuss later in this work, a hard and fast definition of "islands" is particularly difficult to anchor. Therefore, I once again offer my own application of the idea. And even this notion becomes prickly as I ask if islands are an idea or an actuality. Are they landscapes or ideoscapes? Ontological mirrors (as in Foucault's concept of the *heterotopia* [1986]; see Chapter 9) or deeply lived realities? They are all of these things and more. Islands are both geographic and ideological, and their hybridity and division in many realms is the core focus of this book.

For me, and in my work, islands are bounded realities, and while most often the starting point of my inquiries is a literal piece of land surrounded by water, there soon emerges a profusion of layers and dialogues within these confines; ideas float to the surface and the polyvocal reality of these places makes itself known. Along with this fluid definition, it is important to acknowledge the lived realities of the inhabitants of these places and to recognize that each island and islander maintains a valuable and relevant view of their place in the world.

## How to Use This Book: Intentions and Advice

I wanted to write this book for anyone, not just for islomaniacs, anthropologists, and wanderlusters. Perhaps it will be useful in teaching anthropology, or maybe it can serve as an ad hoc travel guide for an ideological or practical archipelagic journey. I can also imagine it coming to rest in the hands of someone wondering what ethnography is, and why it might be relevant to their life. Above all, this work should be seen as a series of points of light, dotting the night sky, opening tiny windows of reflection that lead to larger spaces of inquiry. It is a collection of collections, and an assembly of waking ethnographic dreams. From a less impressionistic perspective, I want this work to make anthropological thinking and writing accessible and relatable; I want readers to see the magic of everyday life through my engagement with the people, places, and things of islands. As I often tell my students, once you've learned how to think anthropologically, there is no going back.

The mini-essays that follow can certainly be read in the order that they appear, but they can also be read as standalone pieces. Readers can island-hop or follow the flow as they see fit. In the end, I am simply encouraging people to consider how and why islands have (and continue to) become present, persistent, and resonant subjects of anthropological (and popular) attention. Hopefully what follows will offer an anthropological way of seeing and thinking about the world through the lens of islands.

## A Note on My Particular Positionality

As anthropologists, it is important to recognize that we undertake our individual forms of analysis using a particular set of lenses, and while these lenses are carefully considered they are also formed and informed by our positional histories, psychologies, cultures, and languages. In what follows, I have done my best to acknowledge and incorporate other perspectives and place them in conversation with my own.

And at the same time, I think it is useful to also acknowledge that anthropology and writing are both inherently positional pursuits, premised on the unique viewpoints of the author. I cannot answer every question from every perspective, nor is that my intent. I cannot offer readers a definitive truth about anthropology and/of islands, but I can present an honest, reflexive, and rigorous set of insights that I hope will encourage readers to attune and reframe their thinking on the subject.

So, please forgive me if I miss a step you were expecting me to take. I ask you to accept my ideas as thoughts in motion. This book represents a snapshot in a moment, but I hope it will provide some illumination and add a new set of applications to your toolkit as you consider the intersections between islands and anthropology.

Justin Armstrong

# Acknowledgements

Thank you to all of the people, places, and things that have made the researching and writing of this book possible. Specifically, I'd like to thank my supportive and patient wife, Heather, who has always encouraged me to dig deeply and consider carefully. Thank you to Anne Brydon for first bringing me to Iceland and for your friendship and insight over the years. Thank you to my advisors, friends, and colleagues including Petra Rethmann, Allen Shelton, Tim Kaposy, Andrew Pendakis, Justin Sully, Susan Ellison, Adam Van Arsdale, and Ann Velenchik. I would also like to thank Elísabet Gunnarsdóttir for offering a miniature writing residency at Arts Iceland in Ísafjörður, where I was able to finish this book. And thank you to my wonderful Wellesley College students who continue to teach me about the magic of everyday life.

This work has been funded in part by The Marion and Jasper Whiting Foundation, The Knapp Social Science Faculty Research Grant at Wellesley College, The Social Science Research Council of Canada, The Saskatchewan Heritage Foundation, and Faculty Awards at Wellesley College.

# 1 Arriving

## An Introduction to Island Anthropology

### Confessions of a Mainlander

In all of their myriad forms, islands have always called to me. Sometimes gently, but more often than not, with a deep, old gravity. Perhaps my clear awareness of this beacon has something to do with having been born in the landlocked Canadian province of Saskatchewan, distant from any watery boundaries. Islands always seemed like beautiful and rare imaginaries, places that I knew were real yet felt wildly distant. Like passionfruit.

In this world of now, every person, place, and thing is always-already a bit of an ideological island. As humans, we often find ourselves compelled by our counterpoint, drawn to the vacuum of our experience and understanding. For me, the edges and borders of islands seemed somehow reassuring and defined, a contained unit of place so unlike the endless vastness of the prairie. Or as Edmond and Smith (2003:2) have claimed, "[b]oundedness makes islands graspable, able to be held in the mind's eye and imagined as places of possibility and promise".

I am certainly aware that this interest represents what could be called a *mainland gaze* toward the island Other, a fetishized essentialization and isolationist imaginary that sees the island as an object of curious inquiry, rather than one of lived experiences for its inhabitants. In what follows, I do my best to navigate this difficult passageway in the hope that I might carve out a way of engaging with islands and islanders on their own terms (Godfrey 2008), in a way that includes a variety of island perspectives and voices. Still, this mainland gaze is also the embodiment of my enculturated view of islands that has seen them as "islands in the sea" instead of "a sea of islands", as Tongan anthropologist Epeli Hau'ofa (1994:153) has so aptly described this outsider view of Oceania. Here, Hau'ofa forces a shift in perspective, illuminating the conflicting *etic* (outsider) and *emic* (insider) understandings of islands as discrete, separate, and bounded, and islands as an interconnected world where water and land are not always clearly

DOI: 10.4324/9781003297581-1

defined borders. To imagine islands as perfectly isolated and self-regulating is to discount the myriad of political, social, cultural, and geophysical connections that truly situate them as ideological and ontological networks, unbounded and intimate. To return to Edmond and Smith (2003:5), "[i]slands are not pure: they are subject to breaching and incursion, both natural and cultural".

Alongside this aim of unpacking the cultural relativity of defining islands, and perhaps rather selfishly, I frame my experiences and impressions on, of, and with islands as a way of considering the multiplicity of cultural connections, interventions, and disruptions that we now find on Earth, itself an island adrift in the blackness, a pinprick of life in a lonesome sea. Maybe, in a small way, everyone and no one is an island(er).

My affinity for islands is what eventually led me to study, teach, and practice anthropology, which in turn led me back to islands. And this is where I find myself now: considering the intersection of a landform/waterform, cultural dynamics, and a discipline premised on translating *away* into *home*. Anthropology has given me the framework and tools to begin the lifelong endeavour to consider my attachment to islands, and to draw out the lines that connect my anthropological and writerly imagination to these very real places. As my particular anthropological view began to solidify in graduate school, I realized that this interest was not limited to geographic islands, but also to remote and isolated places bounded (and bisected) by forces other than water, including politics, ideology, infrastructure, and landscape. Now my interior archipelagos hold both terrestrial/geographic and ideological/metaphoric islands. My intent here is to develop a map of (and to) a world that exists in a global reality as well as in locally imagined and lived geographies and topographies.

In my anthropological world, islands are both fetish objects (in that they maintain certain "magical" qualities beyond their essential nature) and ethnographic subjects. They populate my anthropological and touristic imagination and form my movements between and across landscapes. I am not an islander, and I make no claims of objective knowledge or definitive truth about their people, places, and things. Following James Clifford (2020), I can only offer a *partial truth* of islands, a collage of reflections, impressions, and insights. This splintered fragment of a half-truth contains the experiences and encounters that I have accumulated over many years of engagement with island(er)s in an anthropological context. Here, I am fully aware of my continued mainland gaze, and how this view forms a unique positionality that I continually manage and mediate. I hope to continue what Édouard Glissant (1997) has described as the "poetics of relation" as a means to trace the beautiful, difficult, and illuminating connections and breakages as I employ an anthropological toolkit to unpack some of the key cultural resonance of islands within (and outside of) the contemporary global ebbs and flows, frictions (Tsing 2005) and erasures (Augé 2004) of culture, time, and place in this constellation of island connections and intersections.

I think that we are all fascinated with islands to some degree because they offer a kind of safely distant screen on which to project our own private dreamworlds. They present a speculative past, present, and future where we can quietly wonder, a heterotopian (Foucault 1986) antidote to our ordinary mainland lives. Whether those dreamscapes are made of swaying palms, wild solitude, or unknown and exotic cultural forms, when viewed from the mainland, islands become a *tabula rasa* for thinking away. This notion of *thinking away* is one of the key concepts that I work to unpack in this book, exploring the ways that we use real and imagined distance, isolation, and remoteness to better understand the close-to-homeness of our everyday experience. *Thinking away* allows us to wonder at a distance, to believe in the potential for bounded meanings and carefully constrained lives framed by water and/or other forces.

As John Gillis (2009) outlines in the introduction to *Islands of the Mind*, a fascination and fetish for islands is rarely held by the residents of these places. Islomania is an affliction of the mainland mind. For the mainlander, the island serves as a gently out-of-focus location for yearning and a mis-remembered nostalgia for the future, the way life could have been (see Chapter 9). And still, it is important to remind ourselves that islands are not simply foils for our speculation, fantasy, and academic inquiry. They are real places occupied by real inhabitants with real lives. To my mind, this dynamic interaction between mainland and island, islander and visitor, and imagination and reality is precisely where we can gain the most productive awareness of the value in *thinking away*.

I will be the first to admit that I hold a deep and abiding infatuation with islands. A curious affinity, a psychic tether. And perhaps I have developed a relationship with islands to a level that is often called "over-rapport" in anthropology, where we become too familiar with our subjects and begin to lose our prescribed sense of scholarly distance. But, as I often tell my students, we cannot (and should not) completely rid ourselves of bias and proclivities, and therefore we must acknowledge and manage them in the hope that we might better understand our relationship to the people, places, and things that we study. This project is about embracing, defining, and reconfiguring my connections (and disconnections) with island(er)s.

I am not beholden to one particular island or island culture; my attention to islands, both real and metaphoric, rests in their ability to present models for thinking and being in the world. Here, islands are useful and necessary modes of *thinking away*, *through*, and *with*. They are ontological capsules that convey new and exciting angles of vision. I love islands for their fluidity, assemblage, and potential. Limits are important for generating ideas, and the island, as both landmass and ideological construction, provides a perfectly beautiful constraint for *thinking away*.

## A Very Brief History of Anthropology and/of Islands

> "Imagine yourself suddenly set down surrounded by all your gear, alone on a tropical beach close to a native village while the launch or dinghy which has brought you sails away out of sight."
>
> —Malinowski, *Argonauts of the Western Pacific*, 1922

In two realities, pioneering anthropologist Bronislaw Malinowski and I stand together, looking out over our respective oceans. Other places, other times. He watches the ship disappear over the Trobriand curve in the Earth as I squint to catch the last pinpoint of the tiny propeller plane as it evaporates into the Atlantic fog off Maine's Matinicus Isle. We turn to face our work, the (un)knowable and (un)bounded realms of the island.

Modern anthropology began on islands. Its history is forever entangled with these places. Islands served as rocky proving grounds, clandestine laboratories, and ideological wake-up calls for the discipline by offering a place to disassemble and reassemble the human "instrument of knowing" (Ortner 2006). Islands become us. Similarly, our work continues as its own kind of island: sometimes real, sometimes as figurative islands of our academic and writerly attention that form geographies of our ethnographic focus. As anthropologists, we need limits in which to effectively translate *there* into *here*. We need borders on our imaginings. Islands are the arrival and departure for anthropological approaches to people, places, and things. Still, little has been written on the direct intersection of islands and the pursuit of anthropological understandings, especially considering the pivotal role that islands have played as stages and subjects from the beginning of the ethnographic endeavour.

Early in the last century, Malinowski arrives in the Trobriands to find a test site, a field lab that will allow anthropology to step out of its armchairs and off of its verandas. This island group became the setting for many of the foundational anthropological discussions of exchange, sexuality, and supernaturalism (see Malinowski 1922, 2003, and 1992 respectively). The islands and their inhabitants framed his questions and furnished his answers; they offered a relatively bounded reality out of which emerged practical field anthropology. During his fieldwork, Malinowski embeds himself in this world, he marinates in the everyday lives of the people and participates in their daily routines, a task made much more manageable and comprehensive by the islandness of his chosen site. And I will be the first to admit that Malinowski's work is not without significant dilemmas and colonial hangovers (see Malinowski 1989), but nonetheless, it is an important piece of the formative puzzle of anthropology on/of islands. Similarly, A.R. Radcliffe-Brown's (2013) seminal work in the Andaman Islands examined another compact location and provided the basis for the formulation of the core tenets of anthropology's approach to functionalism, along with a broad-ranging description and analysis of social, cultural, material, and supernatural beliefs of the islanders. The Andaman group is also the location of North Sentinel Island, perhaps the most challenging island for the anthropological imagination. The inhabitants of this remote island violently reject any form of outside contact and as a result the details of their culture are virtually unknown beyond its shores. This place represents perhaps the last *terra incognita* of islandic imaginations, both anthropological and popular. The idea of an unknowable island forms a perfect mythology of the "lost tribe", a trope that when combined with the most unreachable island in the world becomes almost irresistible. And it is here that we see the human fascination and innate curiosity for the unreachable, unknowable distance of the island reach its height.

After Malinowski, Annette Wiener finds herself on the Trobriands, set to revise and reframe his viewpoints on *kula*, the ritualized exchange system *par excellence* in economic anthropology. This system truly reflects an archipelagic understanding of cultural forms that could only emerge out of island cultures. Margaret Mead, arguably the most well-known cultural anthropologist outside of the discipline, also conducted her ethnographic fieldwork on the island of Ta'ū in American Samoa, yielding some of the most significant discussions of the anthropology of gender and sexuality. Marshall Sahlins' work in Polynesia in the realms of historical anthropology offers a number of productive dialogues on colonial legacies, particularly in *Islands of History* (1985) and *Anahulu* (1992).

Anthropologists still seek their subjects in and on islands: large and small, near and far, known and unknown. The history of what we do is inevitably entangled with these places, and in many ways, it can be said that all anthropological projects are (about) islands. Here, it is important to remember that in thinking about, through, and with islands, we must temper our curiosity with relativity. We must remember that the island is both a focus and a locus (Ronsrtöm 2013) of our contemporary and historical anthropological attentions, and at the same time, it is also home to real people, places, and things that deserve our respect and thoughtful engagement (for more on this dynamic see Godfrey 2008).

## Literary Islands

Islands have always been a part of the stories we tell ourselves about ourselves. From Homer's Ithaca to *Lost*, we have always been fascinated by the narrative potential of island places. In many of the same ways that islands have become a focal point for anthropology, they have also served as fertile ground for literary endeavours. In this capacity, islands function as a clearly defined distance, a place easily imagined as remote and extraordinary. They are locations quickly discernible as existing outside the flows of everyday life. While there are innumerable examples of islands as both characters and settings in all forms of art, a few immediately spring to mind as examples of how islands serve as vehicles for essential reflections on the human condition.

Alongside the classic island novels *Robinson Crusoe* (1719), *Treasure Island* (1883), and *The Island of Dr. Moreau* (1896) sits Aldous Huxley's 1962 novel, *Island*, a text that makes a clear case for the resonance of utopian island imaginaries and their inherent, seemingly inevitable fragility. This is the recurring vision of islands that I see as *precarious paradise*, a place of deeply contingent perfection. A more sinister vision can be found in the 1967 British television series, *The Prisoner*, which presents an island-as-prison motif that uses a village bound by mountains on three sides and the ocean on the fourth to form a virtual yet inescapable island. Here we also find the theme of the unknowable island wherein the central character, known only as Number Six, never finds out how or why he has been imprisoned, nor does he understand the true boundaries of his prison. Along similar lines, J.G. Ballard's *Concrete Island* (1974) examines the island as another kind of prison, this time as a site of entrapment on a roadway median set amongst a swirling sea of cars, a definite parable for human imprisonment by our reliance on technology. Another cautionary tale can be found in the fictional Hebridean island of Summerisle, which serves as the backdrop for a fetishized vision of rural paganism in the 1973 folk horror film, *The Wicker Man*. Here the island becomes the container for the mystical and dangerous Other, a warning about transgressing the boundaries between *here* and *there*.

In these examples, the island can be understood as what Judith Schalansky (2010) has described as both paradise and hell. Just as in anthropological discourse, the island becomes a medium on which to pose and address questions of being. It forms a bounded framework with defined limitations that offers microscopic views that allow audiences to *think away*. The prevalence and persistence of these narratives certainly signals the cultural significance of islands in contemporary imaginations of place. Here again we see that islands are good to think with.

## How Islands Are Good to Think With

Real or imagined, islands form a unique canvas on which we paint romantic, academic, touristic, and utopian dreamscapes. Anthropologists are certainly not immune from these imaginations, and we often employ islands as literal and/or figurative sites for our inquiries. From the particular perspective of structural anthropology, islands can be seen as totems (Levi-Strauss 1974), in that they provide a kind of connective tissue between abstract ideas and lived realities. They form an analytic conduit to a new cultural awareness. Islands are "good to think with" because they provide a concrete focal point for individual and collective senses of place, a fascination with the unknown and isolated, and the gentle production of a seemingly unproblematic Other (of course, this production is always-already problematic in some way). For anthropologists, islands are useful "to think with" in that they present a bounded cultural entity, often complete with limited outside influence and unique, observable forms, a view similar to MacArthur and Wilson's (2001) concept of island biogeography wherein the isolation and remoteness of islands produces unique forms of non-human animal and plant life. Of course, it is important to remember that even on islands, culture cannot (and should not) be distilled into a singularity. Islands, regardless of their size, remain internally culturally diverse and variable, as well as being inherently connected to global flows of ideas and things. Here, I am reminded of a research trip to the small island of Fogo (see Chapter 4) off Newfoundland's northeast coast, where despite occupying just over 250 square kilometres, its nine villages maintain distinctly different accents, architecture, and ethnic origins. Therefore, we cannot speak of a uniform Fogo Island culture anchored in place, but one that exists in the dynamic medium of globalization and cultural change. Islands are good to think with, but we must be critical and reflexive of our tendencies to see them as perfect containers for bounded cultural forms.

John Gillis (2009) reflects on the significance of islands as vehicles for both imagined and real forces in human understandings of the world, specifically drawing out the concept of thinking with islands as what anthropologist Sherry Ortner (1973) has described as key symbols of thought. Here, islands become *root metaphors* that allow cultures to anchor and conceptualize their worldview through symbols. Consider the vast number of books, television programs, and films that hold the island as a key symbol for a world apart, a place that exists as bounded and removed from the temporal and ideological flows of everyday life. As Ortner outlines in her essay, anthropologists are always looking for symbols in a given culture which appear to have a pervasive resonance and significance with group members. Certainly, islands fit that mold in a number of global cultural contexts.

The duality of the island/mainland relationship is also important to consider here. Again, taking Levi-Strauss' (1983) work in hand, we might consider a kind of *island dialectics* wherein the island/mainland structure is synthesized into a thinkspace that occupies neither location. Islands are good to think with because they present a means of pulling the mainland out to sea while simultaneously moving the island outside of its imagined boundedness. The island that is *thought with* becomes a conceptual bridge, an abstracted ontology for experimenting with understandings and relationships between the island and mainland.

The purpose of this book is not to unravel and unpack the meaning of islands, or to provide a definitive text on island anthropology; rather it is here to consider anthropology through and with islands. To understand islands as collaborators and co-authors here is particularly important. To think with island(er)s means to engage with them in conversation, to think alongside them, as well as to employ them as locations for critical ethnographic inquiry. This dialogic way of understanding islands allows us to consider their agency and their ability to produce their own kind of ontology. What might be termed an *island-oriented ontology* can provide a useful intervention into how islands produce distinct cultural Selves. How can we begin to understand a way of seeing from the island outward? Similarly, perhaps we should be considering islands as a collection of planes of immanence (Deleuze and Guattari 1988), an assemblage or constellation of layers (see Chapter 9). As cultural, economic, touristic, historical, biological, geological, cartographic, and aesthetic realms existing in one bounded location, islands can be viewed as modes of perpetual becoming, always shifting their internal and external meaning.

## Assembling a Toolkit

Reflecting on how islands can be good to think with, about, and around, it is important to consider how we can begin to assemble an ideological, analytical, and experiential toolkit to help form our future encounters with islands. In teaching both writing and anthropology, I encourage my students to develop toolkits that will form and inform their interactions with their subjects. Here, we consider the relevant phenomenological, ontological, and tactile modes of engagement, along with appropriate modes of inscription (writing, image making, sound recording, etc.) and ethical apparatuses. Each island requires a unique version of our individual and disciplinary toolkits. A cultural geographer, a geologist, and a historian will all hold specific (yet often intersecting) tools in their kit and apply them in ways relevant to their individual and disciplinary approach.

As an anthropologist interested in anthropologies of exchange on the Pacific island of Yap in Micronesia, my toolkit contains approaches, theories, and methods derived from my disciplinary background as well as my ontological outlook. Structured interviews and survey data are not part of my particular toolkit, but Geertz's (1998) practice of "deep hanging out" and Marx's (1967) notion of commodity fetishism are. As writers, anthropologists, and humans generally interested in the world, we constantly assemble, revise, and reassemble our toolkits. Sometimes this development happens subconsciously, but here I make the case for allowing that toolkit to be a bit more clearly defined and purposeful, letting it rise closer to the surface of the everyday, and become just a touch more intentional. It is important to ask how we see the world and how we intend to translate our unique experience of it. We must codify and anchor our approaches to make them our own.

A toolkit for thinking with islands offers unique benefits and particular challenges. From the most practical items (what can you carry? How often can you get necessary supplies?) to deep theory (how does your positionality form and inform your interactions? Where do you fit in? What disciplinary structures and conventions do you need to reconsider and revise?), the toolkit of an island anthropologist must be thoroughly considered and carefully packed. We must choose wisely and understand the resonance and relation between everything we have assembled.

## Cultural Ecotones: An Anthropology of Divisions

Ecologists often speak of ecotones as sharp demarcations between biological communities, such as a clear shift in the types of plant or animal life between two regions. Here, an edge is formed between habitats, clearly marked by a line that can be the result of natural or human forces. In these spaces, species adapt and form new modes of survival (Nemeh and Longe 2021). Similarly, as an anthropologist interested in the potential for studying the cultural significance of mainlands and islands, the notion of an ecotone, or defined edge of culture, is particularly useful to think with.

Arjun Appadurai's (1990) conceptualization of global cultural flows and their overlaps and intersections draws this notion of ecotones into more direct contact with anthropological thought. Here, Appadurai considers how culture takes shape at these dividing lines, how interactions in border zones form significant disjunctures that provide fodder for considering how and why globalization takes no prisoners. The stark boundaries of ecotones serve as a reminder that everything is not flow, and that network connections often bypass the periphery.

These global flows of capital, people, and information have erased many of the cultural, geographic, and temporal boundaries that had been taken as social facts for much of our history. Benedict Anderson's (2006) concept of print capitalism took the printed word as the primary mode of this shift, but now internet technologies have broadened and accelerated the establishment of McLuhan's (1992) global village. Yet the ecotone that persists between mainland and island remains an undeniable factor in disrupting and limiting these flows. Of course, I am not claiming that islands (geographical or ideological) are exempt from these flows, rather that their boundaries and edges become more clearly apparent in a world deeply infatuated with interconnection. And while these places have been enveloped by many of the same aspects of print/data capitalism, they also become more isolated and less integrated into this system of flows. Take for example the now-abandoned (or in governmental parlance, "resettled") outport of Grand Bruit, Newfoundland (see Chapter 4), a place with a population that was swept up in a series of finance-, ethno-, media-, ideo-, and techno-scapes (Appadurai 1990), resulting in a disintegration of pre-existing cultural patterns and ways of being. Now the ecotone of this place is defined by both its separation and emptiness. The ferry no longer runs to Grand Bruit; its absence is a cultural ecotone.

These borders, edges, and divisions are a prime target for the ethnographic eye since they offer a rupture (Stewart 2007) in the flow. These lines across and along (Ingold 2007) define a question of culture, signalling that there are forms to be mapped and cultural grammars to be defined.

## Betwixt and Between: On Ferries

Ferries are the liminal spaces between islands and mainlands. They are the conduit, the becoming-island, the ties between *here* and *there*. Marc Augé (2009) might see the ferry as a species of non-place, constantly departing and arriving but never really inhabiting a defined location. The ferry to islands is also the site of happenstance and unplanned interactions. It is a generative place where islanders re-encounter one another and visitors build tiny silent stories about one another from across the deck. Ferries present themselves as movable islands to and from, ideological and cultural airlocks that introduce and conclude. There is a strange set of constellations and intersections that takes place on the ferry over. Days and weeks later visitors are reminded of their crossing cohort when they encounter one another at a bar, along the beach, or in the grocery store aisles on the island.

Ferries take other, non-seafaring, forms as well. A tiny propeller plane from Owl's Head to Matinicus in Maine, a sometimes jet from Guam to Yap (see Chapter 2), or a ride in the bed of a worn-down pick-up from the airstrip to town in Uranium City in northern Saskatchewan. A ferry conveys its passengers and the meaning of the place. It offers the first sense of the place, a portable miniature, complete with local ideologies, symbols, and language.

Ferries also mark the island as unbridged, somewhere requiring a special effort or desire to reach, often signalling a one-way-in, one-way-out world. They are the literal and figurative price of entry to the place. The price paid for transportation is also the translation fee. Here, ferries offer a glossary of ideas, a mode of beginning-to-know. They provide the visitor with a chance to adjust and the islander a means to reintegrate. These conveyances function as what de Certeau (1984) has described as *frontiers* and *bridges*, and as such shifts "the void into a plentitude" and "the in-between into an established place" (127). And as I hope to unpack throughout this book, these delimitations form themselves into narratives.

My tattered map of Saskatchewan indicates a ferry crossing. The gravel road dips down to the muddy river's edge. As instructed by a faded metal sign, I flash my headlights and the ferry makes its slow and short journey from the other side of the river. I am back where I started, floating on a tiny island in an islandless expanse.

# 2   Wave Glossary

## To and from Yap

### Hawai'i

From Boston's sharp January morning to touchdown in a thick tropical night. I already feel the heady myth dream of the anthropologist taking hold, and I already resist it. These tropes of the island anthropologist begin to bubble and seethe, and I see myself stepping into Malinowski's shadow in those Trobriand photographs, a ghostly pith helmet beginning to haunt the back of my head. I feel like I'm entering some sort of magical realist novel where ethnographic progenitors form a gang of filmy spectres, forever in the seat beside me. I often feel like these characters exist in a strange state of limbo, between reverie and damnation. I can't ignore them, but I also need to be conscious of my role here. Wheels down, lights up, and soon I'm swallowed whole by the fluorescent humid evening.

An open-air terminal filled with evening's mellow bustle. Honeymooners and wayward locals billow out into the shorts-and-sandals dark. It's even more imaginary than I'd imagined, like meeting a wealthy aunt who sends sporadic Christmas cards from New Mexico. From the backseat of an over-priced taxi, a theme park dreamworld unrolls as we pass the shipping containers huddled under highway overpasses. And soon, Waikiki Beach oozes and drips with heavy syrup; all the same money seeps from corners and slips out under sliding doors. I had never been to Hawai'i, always writing it off, always pushing its presence aside in my traveller-anthropologist imagination. Because of course I was a traveller, a tourist even, before I was an ethnographer. Still, my ethnographic eye is always open, even if it's shaded behind sunglasses and a baseball cap. Even if this place is only a waypoint along the line.

This island is pre-programmed in our thinking; we absorb its meaning through osmosis. Nobody has to tell us what it means; its meaning is its existence. Hawai'i is always-already part of our collective imaginations, a distillation of the tropical idyll escape, like Tahiti but less foreign. It evokes

DOI: 10.4324/9781003297581-2

itself in a strange feedback loop of an unwakeable dream-state. Every "local" I meet tells me that they came for a visit and never left, becoming a sort of self-fulfilling wish, comforted in their collective disavowal of the mainland. It's like they've been let in on a secret that one can live a vacation, and in some ways this thought is really only resonant because of Hawai'i's islandness.

And again, it is important to remember that this is the outsider view, the etic perspective of Hawai'i that sees it as a discrete and separate place, an "island in the sea". But it is also (and has been) "a sea of islands" (Hau'ofa 1994:153) with complex connections and everyday people, places and things without exoticism or vacation glazing. Like many "destination" islands, there is a layer of banality and extra fees to have Amazon packages delivered. As Jamaica Kincaid so incisively notes in the opening chapter of *A Small Place* (1989):

> . . . every native of every place is a potential tourist, and every tourist is a native of somewhere. Every native everywhere lives a life of overwhelming and crushing banality and boredom and desperation and depression, and every deed, good and bad, is an attempt to forget this. Every native would like to find a way out, every native would like a rest, every native would like a tour. But some natives—most natives in the world—cannot go anywhere.
>
> (18)

A key element of a genuine and productive anthropology on/of islands involves awareness and critical reflection on the dual realities of islands: the imagined island and the lived island. And just as tourists imagine and fetishize islands, so do anthropologists. We imagine them not as perfect touristic escapes, but as perfect ethnographic laboratories, and we might be well-advised to reconsider this approach, or at the very least be conscious of our viewpoints and positions. Of course, this is not to say that we always-already understand these places in these ways, but we must reckon with our histories and motives and work to push back against some of the proclivities our discipline is subject to.

## Yap

Yap is a *high island*. All of the states in Micronesia (which includes Yap, Chuuk, Pohnpei, and Kosrae) have high islands, the landmasses that rise above the ocean and form political and social hierarchies. Here, the topography of islands becomes a form of governmental influence. In Yap, power radiates eastward to the outer islands including Ulithi, Sorol, Eauripik, Woleai, and Satawal. As subjects of Yapese high island politics, the outer islands have formed their own tribute and trade networks with Yap, including a specific linkage with Ulithi and the Gagil region of Yap known as *sawei,* which is based on a sort of tenant-landlord tribute relationship (Lessa 1950). To my mind, these forms of political and economic interconnection reassert the importance of considering Hau'ofa's (1994) concept of "a sea of islands". Despite the almost 500 kilometres between Yap and Satawal at the far eastern edge of the state, these islands form a unified whole, a constellation of connections, an *aquapelago* (Hayward 2012) of culture across water.

I came to Yap to study the unique form of currency, the "stone money" known as *rai.* This medium of exchange has captured the imaginations of economists and anthropologists for years, beginning with William Henry Furness III's 1910 *The Island of Stone Money* and moving toward Milton Friedman's 1994 nod to the *rai* in *Money Mischief,* and finally taking a distinctly archaeological turn with the work of Scott M. Fitzpatrick and Jennifer Pinkowski (2020). For me, *rai* had surfaced briefly in an undergraduate course and again in my teaching of introductory anthropology but was never more than a flickering moment of notice. As my interest in economic anthropology and "thing theory" (Brown 2001) developed from a variety of angles, I found myself teaching a senior seminar on anthropological exchange which featured a class meeting dedicated to the stone money, which renewed my interest in this fascinating subject. Two years later, in 2020, I received a Whiting Fellowship to visit the island and conduct some preliminary research for a new project on the cultural histories of money. As with all best-laid anthropological endeavours, my focus shifted almost immediately on arrival. Of course I was still interested in the *rai,* but I quickly learned that there were other compelling and lesser-known forms of exchange, including the *sawei* system mentioned earlier.

Touching down on Yap in the middle of the night, I flowed with the other passengers into another open-air terminal, much smaller and more chaotic than Honolulu's. It seemed more honest, lived in. A series of tiny whirlpools and eddies brought me between customs and baggage claim as I passed tattered posters warning about fire ants and dengue fever. Luggage was hurled and piled from a gaping doorway without any attention to order. Empty Rubbermaid totes covered in dogeared strands of packing tape bumped and tumbled across cardboard boxes, backpacks, and misguided rolling suitcases. Teenage boys with glittering earrings and chains grabbed and stacked the plastic bins with fluid ease, calling across the tiny airport in Yapese and English. The Betel Nut Boys were back in town. I would later come to learn that these kids were ferrying the bins of betel nut, a mild stimulant, to Guam to be sold. Betel nut became a constant theme in my time on the island; it was prevalent everywhere I went and served as a mode of introduction during a meeting with the chief of Ulithi. Here, another form of exchange emerged, one that intertwined these island places to create yet another bridge between Yap and its neighbours.

Four wide-eyed Mormons in their white short-sleeved button-ups and name badges bob through the airport, surveying their new world. Their landing cards listed their purpose as "missionary", and I am reminded of another kind of interconnection and the impositions of visitors. In their attention to the island, the missionaries were not tourists or anthropologists; they occupied another imaginary, that of the island as a godless periphery. And again, another outside layer is laid across the low hills of the high island.

The forested parts of Yap are jungle-*like*, not the television personalities that I'd encountered in places like Colombia and Panama; here they are a gentle tangle of damp undergrowth turned inside out. Dirt paths and rutted lanes splinter off the main roads, winding toward ad hoc structures for everyday cooking and sleeping. Nothing precious, nothing overwrought. Alongside these trafficked lines are the careful and beautiful stone roads, so thoughtful and purposeful, they seem to have existed before the forests in which they embed themselves. These routes form a network of pre-asphalt connections between villages on the island, perhaps another kind of ecotone, a series of generative conveyances through the quiet forests. And still another network parallels these roadways: the ghost paths that also trace themselves across a spiritual geography of Yap (see Chapter 9). I was told by several islanders that these routes allow ancestor spirits to navigate another version of the island, and here is another layer, another meaning of the island. This understanding of a mirrored spectral landscape echoes many of the same themes of haunted geography that I encountered in Iceland and the Faroe Islands. These views suggest a potential for another kind of ghostly connection outside of the discrete islands.

An uncanniness accompanies the discovery of downed Japanese fighter planes and rusted-out artillery. Echoes of other times. This island, like all islands, is layered deeper than it is wide. There are only limited realms of outward expansion, and the histories, narratives, and dreamworlds pile themselves thickly. Stone money sits in view of an abandoned Chinese fishing boat that was set adrift and landed on Yap several years ago. Hollowed-out plane carcasses teem with vines, a smudged-out sign written in English marking the spot in time.

## Rumung

Rumung is a forbidden island. Yap itself is made up of four smaller islands separated by narrow canals, except for Rumung, which can only be reached by private boat. It has no shops, schools, or electrical grid. Some of the largest *rai* are on this island's island. The veil is thin here and I sense a dense quiet hiding in the forests. I've encountered this veil, a sense of place that haunts and hovers just beyond several times before, most strongly in Iceland, but also in Wyoming and Saskatchewan. Not supernatural, more extranatural; an unplaceable gauzy curtain hiding something and uncertain about my presence.

As the small motorboat rounds a final point of land, a men's house appears at the base of the cove; a warped airplane propeller and a rusted-out engine block from a Japanese fighter plane sit propped up against the front of the structure, looking out to sea and marking another era of colonialism and one more layer of an island no longer isolated from historical imaginations. Nobody can visit the island without being accompanied by a local. My hired chaperone, S, barely speaks. His Tapout t-shirt and basketball shorts seem disjunctive, but I don't know what I was expecting. He tells me in a hush not to photograph one of the large *rai* because someone died bringing it back to Yap. It is dishonourable to hold these images. He tells me that the *rai* have only been allowed to be photographed since 2015. Islands become preserves, sometimes through will, sometimes through apathy.

## Ulithi

Ulithi is an outer island of Yap State, the periphery to a place already absent from most people's maps. A dot of land amongst the thousands floating on the wrinkled surfaces of classroom globes. This place is one of those pinpoints sprinkled across the South Pacific, even more distant in flows of imaginations than Yap, which oversees this atoll about 200 kilometres to the east.

Visitors to Ulithi need to obtain special permission to travel there, so on the day before I was scheduled to fly over, I had to visit the Ulithian council to get approval. This exercise held a new tone for me. As an anthropologist primarily working on the topics of ghost towns and Northern Atlantic islands, I'd felt somewhat removed (insulated?) from the sticky postcolonial sense embedded in Oceanic ethnography, but here I was, deeply embedded in threadbare tropes. Thinking back to teaching my anthropological methods course, I tried to take my own advice and remain critical and reflexive of my position, but the other part of my ethnographic brain left me feeling ideologically dizzy about my role. In some ways I felt like I was (re)playing the role of The Anthropologist from one of Malinowski's Trobriand photos. Bronislaw was becoming a much more constant spectre on the island and in my imagination.

The Council is located in a low cement building a few miles outside of Yap's capital, Colonia, and on this typically hot afternoon in January, I arrived at the front door, took my shoes off, and stepped into the cooler darkness of this somewhat makeshift government building. A stern-looking islander sat behind one of the two desks occupying the large open room, eating his lunch from a plastic container. Silence quickly filled the room. After all my years as an anthropologist, I never get used to these fish-out-of-water moments, and again I told myself to heed the advice I always offer my thesis students as they head out to start fieldwork: ride the wave, get freaked out, and embrace discomfort. The man, now still in the midst of his meal, looked thoughtfully and quizzically at me, or at least in my direction. I stammered and stuttered that I was hoping to visit Ulithi and I was here to get the proper forms signed. He told me to sit on a folding chair in front of the other desk and wandered off through a door at the back of the room that I hadn't noticed until now. A few minutes later another man emerged, sat down at the desk, and began rummaging through the desk for the paperwork. The first man returned to his perch to continue his lunch, constantly glancing at me over the edge of the container held close to his mouth, a little more than curious about my interest in Ulithi.

After a number of questions about my intentions for visiting the atoll, I was granted permission to travel to Ulithi, handed an official letter, and told that I should bring betel nut and 20 dollars for the chief who would be hosting me. The next morning, I found myself back at the airport, being weighed along with my bag to make sure that the tiny mid-century plane would be able to accommodate the combined weight of its passengers and cargo. And before long I was up in the air, sifting through billowed clouds with a tall Midwestern transplant at the controls. We chatted over the headsets about his life as a missionary pilot in Yap and the importance of these air linkages for the outer islanders. Once again, unfurling a more clearly defined aquapelago (Hayward 2012) of practical and ideological constellations. And as I have been so many times before, I was reminded that islands are made by connections, not separations.

Before long, Ulithi whispers out of the ocean at a kind of half-light horizon, its 40 islets slowly taking shape under the wings. With less than five square kilometres of land spread around one of the largest lagoons in the world, this place is a wilderness of being-on-the-edge. In the moment that the wheels hit the tarmac, it is a dreamworld made instantly real.

The pilot tells me that he'll be back to pick me up after making a trip even further afield to some of the other outer islands, and he passes me off to a local man who offers to take me to visit the chief and show me around the island. And once more, I watch my plane vanish over the treetops as I become another island in a new sea.

I spend the day being shown around in a struggling pick-up, using the WWII-era roadways that weave their ruts through the low jungle to connect various nodes of island being: the high school, the chief's house, the abandoned fishing resort, my guide's house, the post office. More lines, more layers, more constellations. And I worry about how to translate this untranslatable moment, how to anchor this experience in some kind of meaning. How can I write these people, places, and things into being? How can I put the world in context?

This place is a collection of islands so far-flung that they barely register, yet they are neither minor nor outlying. Their resonance is just as relevant, their connections are just as strong and fragile, and their ways of knowing are just as important. And perhaps the lesson that I can hope to convey from my time in Ulithi is that both island and anthropological forms of knowing can help us to connect, integrate, and illuminate our disparate world. Hayward (2012) has described these intersections as "aquapelagic assemblages", an understanding of islands as more than the sum of their landmasses. Here, the ocean, the islets, and the cultural threads that wrap them together form a unique way of knowing, a way of being in and seeing the world.

All of the nodes of this brief encounter with Ulithi have formed a new kind of aquapelago, an assembled place that is drawn out in lines and layers and connections, taking a new shape that is both geographic and ideological. From my interactions with the Council to my wobbly flight over the ocean to my crisscrossing the island with my ad hoc guide, every point becomes a new connection that shimmers its way into a true "sea of islands" (Hau'ofa 1994). In recognizing and acknowledging the significance of islands in the Anthropocene, we must look for associations and intersections and see islands as being connected and integrated by their boundedness, rather than isolated by it.

As I step onto the plane after a long, humid day, someone from the back of the assembled crowd yells, "Take care of that anthropologist for us, okay?!" A very long line now connects me to this place, and now another one is drawn back there as you read this sentence.

## Antonio B. Won Pat International Airport, Guam

A Cold War haze envelopes another warm night imaginary as the small jet descends on Guam, my first stop en route back to Boston. Always-already a non-place (Augé 2009) woven out of light-without-time hallways. I feel like I know that it's the middle of the night, but it's only a memory of a truth. Marc Augé discusses airports as key examples of non-place, yet he is also careful to note that the marking of a location as such is subjective. One person's non-place is another's locality. From my viewpoint, this airport is a non-placed island, or perhaps it has been mis-placed, waiting patiently for a new integration.

This place has broken free of its moorings, adrift in the non-night as a non-place. It is a kind of island, but one without the connections and intersections. There is no lifeworld here. It is an uninhabited conduit, a passthrough world. This airport evokes only evocation, it is a slipstream fantasy of immobilized moving sidewalks. It is the teleportation device across and over islands, the anti-ferry that obscures travel and ecotone. Guam is a living fossil of colonialism, a geopolitical outer planet orbiting the dying sun of the United States. And then Yap becomes Pluto, a one-time member now set adrift. The sun will implode soon enough.

# 3 On Becoming an Ethnographic Ghost in the Faroe Islands

## Reykjavik to Tórshavn

The domestic airport in Reykjavik serves a handful of destinations in Iceland, along with a few in Greenland and the Faroes. It's relatively free of tourists. Rosy-cheeked families and gangly sports teams bump their generational luggage around the quiet terminal. A few intrepid "travellers" eye each other up from behind expensive sunglasses, wondering whose destination is more remote, who is seeing the "real" Iceland. This little airport now serves as a conduit to the traveller's frontier, bringing the intrepid to their own private dreamworlds. In some ways the airport is more like a ferry terminal, a departure node for remote "islands". Sometimes these destinations are both literal and figurative islands, as in the case of the Faroes: a place apart, the far-flung Other to Iceland's newly minted touristic and financial integration. No longer an Other Europe, Iceland passes that baton to the Faroes as the periphery to a new centre.

On a late afternoon in July, I'm there in the midst of it all. I'm probably an assemblage of all of these characters in different measure. I'm not really anyone to anybody there. I've just wrapped up another round of my ethnographic field course with a group of my Wellesley College students and now I'm off to the Faroes to live in a ghost town at the edge of a fjord for the rest of the summer.

The small plane rolls its way across the nameless North Atlantic. And once again, the wheels touch down on an island, making the imagined real in an instant jolt. Vágar Airport is even smaller than Reykjavik's and seems strangely timeless; even the light quality is hard to anchor in place and time. I collect my bags and hop a shuttle to the capital. As the bus skirts worn-down mountains and speeds through a long undersea tunnel, I strike up a conversation with some university students returning to the islands for summer break after a year in Copenhagen. They tell me that there are more Faroese people in that city than there are in all of the Faroes. Another kind of island, an assemblage of islanders coagulated around identity, a remote island of islanders.

DOI: 10.4324/9781003297581-3

The woman from whom I've rented the house emerges from a neat doorway along one of the tendrils of streets here in Tórshavn. She walks me to a faded hatchback in the shadow of a low apartment block and offers to lead me to the edge of the city in her car. I'd only really driven a manual transmission a few times, but before I can reflect on my inabilities, she's off and I'm grinding my clutch after her as best I can, weaving through roads that were never meant for this. At the edge of town, I fling myself off a roundabout in the direction of Klaksvík, and eventually onward to my destination, Múli. I've been slingshotted out into the unknown, a moving and movable island of my own.

## Tunnels

The Faroes are bullet-ridden with tunnels, some ultra-modern, two-lane roadways that dive miles under the ocean floor to connect islands; others are rough-hewn, unlit passages through the grey-black guts of mountains. Borðoyartunlarnir is a pair of tunnels that link the western and eastern parts of northern Borðoy, the island that is home to Múli. Built in the late 1960s, these tunnels are the second and third oldest in the country. Single-lane (but not one-way) and unlit, the tunnels evoke a sense of liminality. Driving through the narrow passages envelopes and erases; it is the only kind of night in the summer's endless daytime. Tunnels, like ferries and bridges, change the nature of islands. Here, the tunnels begin to remove traces of remoteness; connection begins to evaporate remoteness. *Elsewhere* becomes a little more *here*.

## Múli

At the northern tip of the island of Borðoy rests the abandoned village of Múli. The last settlement in the Faroes to be connected to the power grid in 1970, and one of the last to be connected to the rest of the county by road in 1989. Its last residents left in 1992. Now I'm here in the hope of becoming a temporary ghost, to haunt a haunted place in a tiny way.

Coming here was my attempt at developing a kind of hybrid of method acting and ethnography, an experiment that I continued during a research trip to Hornstrandir in Iceland a few years later (see Chapter 5). One of the questions I'd often posed to myself about my research and writing revolved around the potential for participant-observation (that core principle of ethnographic inquiry) in the absence of people. How could I perform an ethnographic engagement in the spaces left vacant by their human actors? In my earlier research among the ghost towns of the High Plains (see Chapter 8), I'd wondered about the possibility of this sort of hybrid, a collage of an archaeology of the recently abandoned present (see DeSilvey 2006 and Edensor 2005), ethnography, performance art, and storytelling. Would it be possible, in the most microscopic way, to become the final occupant of an abandoned settlement in order to glimpse a shard of understanding of what it might have meant to be the sole resident of a place under erasure? And again, another kind of island emerges from the depths. I have formed my ethnographic Self into a human-as-island, a person enveloped and bounded by vast open space, disconnected from the mainland of my everyday.

In Múli, I spent my days staring across the fjord, writing little notes to myself in my field notebook. Some days I hiked up into the bowl-shaped depression above the village and looked down at an old world, trying, in miniature gestures, to embody the ways of being of people who might have leaned against this same rock hundreds of years ago. I look down at the intricate web of ordinary lives that pressed their feet against the same spongy ground. I work to sit beside myself, to form an etic version of my emic self, a conflicted and constructed being, part anthropologist, part ghost.

And always back to the island. So many strata of lives once lived, so many kinds of islands in one place. The other side of the fjord bears the marks of eons of sediment and eruption, striated rock faces tell a geological legend of accumulation. It's less immediate and apparent in the human counterpart. I'll have to fill in some missing pieces to the best of my abilities. Time builds itself in place, and on islands this seems all the more resonant. Islands on islands on islands, each layer more compacted and opaque. The pressure of history eventually fuses time and place. Here I am reminded of Walter Benjamin's (1968) classic example of the angel of history that watches the accumulation of human time, endlessly deepening, and forever out of reach. Progress, Benjamin reminds us, is inescapable and its ghost-strata are quickly lost and forgotten.

Some days I forget myself. I forget that I am living alone in a ghost town and the end of a fjord. Without other people around, I lose track of my humanness and start to blend into the house, the hills, and the ocean air. Part of it is like a waking dream, especially because it is never truly night here in the summer. Everything becomes a question of islandness; are my responses and sentiments authored by this island? At the headland just north of the settlement, I am easily reminded of the remoteness. The horizon is an uncrossable and undeniable line.

I am reminded of Tim Ingold's (2007) discussion of the cultural relevance of lines as I reflect back on my little window of life in the deserted village. There are lines everywhere here, but most are invisible. They stack and intersect, draw themselves thinly and wrap themselves around their world. They swirl in tighter-yet balls and nests, wound up in their island homes; ever deeper, ever tangled. Ingold tells us that ". . . to study both people and things is to study the lines they are made of" (2007:5), and I would add *places* to Ingold's "people and things", but I agree wholeheartedly with this notion. And if we take this argument to the realm of islands, and specifically this little cluster of buildings, pathways, and crumbling cement, Múli becomes a beautiful and resonant knot of lines cloistered by geography and ideology. Following in Ingold's footsteps, in so many ways, I am an untier of tangles. I lay the lines down beside each other in order to see how they relate and reverberate with one another, and this exercise and its associated description becomes the miniature ethnography of this place. A line brought me here and lines will take me away.

Along with this meditation on lines, knots, and layers, it is important to consider what excites and animates these elements. For de Certeau (1984), this force is the narrative. So once again, the work of the ethnographer in/on the island becomes the practice of assemblage. We animate the lines with our attunements, attentions, and narratives.

## Silences

The silence here is everything. Everything is clung with quiet, and even the sounds are hushed. The car door slams and echoes itself off the sides of the fjord in a weird anachronistic language that tells an emptied story. The ocean makes sounds below the steep cliffs at the edge of the village, but the waves are so far down, and the wind pushes most of the sound back out to sea, headed towards Iceland.

Silence is a hungry ghost. It devours light and shadow and blocks out the sun. Still, it is strangely comforting, another kind of ocean that makes islands of sound. It settles in place and waits for me to fall asleep on the old brocade couch in the living room of the old house. It seems that silence lives longer on islands, its half-life extends in the deep absence of people and things. Of course, this is not true of all islands. I'm sure there are loud islands where sound amplifies and overlaps, unable to escape across the water. Perhaps Ibiza is like that, but here in Múli quiet is the ruling monarch.

I stare into the silence of the fjord that stretches its hush down to the small village of Norðdepil. Everything is still and quiet and I feel the world slow itself. The only things that move are tiny black bird-dots high above the rock faces and a tremble of dry grass at the edge of the road. The island is trying to trick me into thinking I'm dead.

## Mountains as Pyramids

Malinsfjall peers in at me through the kitchen window; a band of sunlight gouges out a path of yellow-green against its general blackness. A pyramid of non-human origin. Again, the island amplifies and focuses, animates and electrifies. Without noise, the voices of these mountains become deafening. I can hear the old rocks speaking extinct languages we've always-already forgotten. This mausoleum of ancient pre-human time watches me write through the salt-stained kitchen window. The sun at midnight gave it eyes to see.

And while I think that my mental stability during this time may have been wavering, I feel that I began to see the layers of an island place delaminate under the gaze of the mountain. And as this awareness surrounded me, I could see the fine details and weird powers of this place; the old wisdom of the island became a clearing in my mind.

For me, this is the quintessential anthropological moment, that slice of attunement that reveals itself in unexpected and purely resonant ways. Being an anthropologist is about waiting, opening, and being. It is about *being there*. The islands—both literal and figurative—discussed in this book have that quality; they allow and invite a distillation of experience, unfettered and untethered. Perhaps this view is too romantic, too fetishistic, too essentialist, but I can only translate what I have experienced, and these experiences in all of their positional messiness are my partial and incomplete truths (Clifford 2020). The island's mountain watches me and interrogates my presence. But this imagined world is just that, right? Still, my response and reflection are ethnography all in itself. The island makes a kind of anthropology and anthropology makes a kind of island.

## Ghosts

As an anthropologist, I think of ghosts as indefinable senses of place. Ghosts are free resonance, a worn tape loop of existence that wows and flutters its way through our consciousness, half-here, half-there, both *home* and *away*.

When people find out that I spend a good deal of my research time occupying and studying "ghost towns", they will inevitably ask if I believe in ghosts and/or if I've had any encounters with them. And this question, like so many in anthropology, is not simple to answer. Perhaps disappointingly, I cannot say that I have had a definitive interaction with a ghost, nor do I generally think they exist, but I can say with utter confidence that I have had innumerable brushes with unknowable senses of place. And like many things, the ghosts of islands take on a uniquely resonant form.

The Faroe Islands haunted me from the moment I arrived. Not as a malevolent force, but as a kind of gentle warning to not become complacent. Instead of trees there are shadows, and the long grey fingers of July's perpetual dusk point and beckon me as I navigate under, around, and across the knives of water between mountains. Here, my immediate sense is that the ghosts are not people, but time and memory embedded in landscape, and as I've often heard, Faroese and Icelandic ghosts can't swim, so they remain bound by the vastness of the ocean along every inch of the almost 700 miles of coastline.

## Time Islands

A boat to an island is time travel, not in the science-fiction sense, but rather a hybrid of time and travel, the moving to, along, and through timelines. The boat passenger becomes keenly aware of the time in distance and reflection required to reach the island. Archipelagos of time, connected and reconnected by more lines that wind and unwind themselves, always at variable rates. Out here, time is another tape left on the dashboard in the sun. It warbles and misses. The beach-bum mythologies of island time are nothing more than colonial views and imaginations of time-out-of-time. The time of these places is deeply rooted and wholly subject to its undeniable passage, but still it is different. How could it not be? Icelanders told me that the Faroes were like Iceland 75 years ago, somehow existing as a half-speed world just beyond the horizon.

And again, like so much, time becomes amplified and more resonant on islands. Perhaps it is more pure, less conflicted and cluttered. Yet, returning to the stratified world of islands, there also exists a set of island times: the local, the visitor, the anthropologist, the geologist, the cartographer, the geological, the non-human. Maybe these layers are more defined and recognizable; maybe they are more defined ways of being in the past, present, and future. Islands, and their times, are not simple and separable; they are made of waves and layers of time that intersect at points to create a new element of the always-shifting aquapelago (Hayward 2012).

# 4 Newfoundland

## A Place Apart

### The Island of Canadian Others

Newfoundland is both part of, and apart from, the rest of Canada. Distant and rocky to most national imaginations, it serves as the topographic and ideological periphery, the outer limit of the country. It is a place that I almost always have to qualify and explain to non-Canadians: off the east coast, joined Canada in 1949, takes at least eight hours to get there on the ferry.

I first came to the island during my dissertation research on isolated and abandoned communities of rural North America, an ethnographic study of places in the process of becoming ghost towns. Initially I had envisioned my trip to Newfoundland as a kind of appendix or counterpoint to my work in the High Plains, perhaps as a type of bookend to the project. What started as an addendum quickly became a deep and abiding affinity for this place. I was tethered to the island by an invisible rope, something that continually draws me back. What was once, for me, barely a marginal outline on classroom maps had become a constant source of inquiry and interest.

Newfoundland serves as an ideological shorthand for remoteness to many Canadians. Distant and missing the *Anne of Green Gables* quaintness that attracts visitors to our other island province, Prince Edward Island, Newfoundland floats hard and rough at the edge of our world. Growing up in northern Ontario, it rarely entered my curiosities, and remained that way well into my adulthood. I knew it existed, but it seemed like a phantom island drawn as a strangely shaped fiction in the corner of the map. Perhaps it wasn't really there, like the grotesque monsters devouring boats near Greenland on old sailing charts.

In many ways Newfoundland is another world, one that is maligned and fetishized. It is a place that sometimes provides a foil for the rest of the country to gain some essential understanding of their identity. It is the quintessence of distance. Newfoundland, like many islands, also exists in its own unique time-space, a fusion of time and place that Bakhtin (1981) has

DOI: 10.4324/9781003297581-4

called the chronotope, wherein time becomes less distinct from place and their meanings and resonance begin to overlap and hybridize. This assembly sees history suffused to landscape and bounded by water. This is not to say that these chronotopes do not exist elsewhere, only that they appear more clearly unclear in island settings, and particularly in my experience of Newfoundland.

My goal in this discussion is not to essentialize, romanticize, or fetishize this place, but I feel that some of the tones and shades that I use to discuss this magically real place may take on these qualities in certain light. Like almost all of the islands I discuss in this book, I am not a local to Newfoundland, nor am I an expert. I am an anthropologist scraping at the surface of place, looking for useful glimmers to share.

## Along the French Coast

Rose Blanche is the literal end of the road. Route 470 terminates just under 50 kilometres after leaving Channel-Port-Aux-Basques, the terminus for the ferry coming over from Nova Scotia. Bumping along decaying almost-asphalt in a dilapidated almost-taxi, I chat with the driver about the dwindling population along the nearly roadless southern coast of the island. The ferry dock where I'll catch the boat to Grand Bruit is empty save for a few trucks that people from the outports leave there for grocery runs or medical appointments. The husk of a fish processing plant leaks woolly insulation and reddish water into the afternoon and a mist-enveloped cemetery stands watch over the little cove. A government sign shows the timetable and route for the ferry that first makes a stop at La Poile before moving on to the end of the line in Grand Bruit, my temporary home for the summer.

Ingold's (2007) lines return as I trace the half-loops on the sign, drawing fragile connections to isolated communities with my finger. Points on a map drifting at the edge of another world. I feel out of place here, a little bit of an intruder. I imagine most anthropologists feel this way when they start a new round of field work.

The much-smaller-than-I'd-expected ferry floats into the harbour, and I throw my bag into a plywood luggage box as I step aboard, somehow sealing my fate when my foot lands; the imagined becomes real in that contact zone. I spend the three-hour trip to Grand Bruit standing on the deck of the ferry, constantly sprayed by the sea and watching the endless rock of the shoreline slip away into gloom and foam.

Since Grand Bruit was resettled in 2010, the coast's lines have shortened dramatically. Now the boat only goes to La Poile, serving a steadily declining population of about 80. Further down the coast, Grand Bruit's inlet is now silent, and past that Burgeo and then on to Francois and Grey River, a couple of outports still hanging onto the coastline, resistant and resilient islands of older time.

## A Contextual Note on Newfoundland's Outport Resettlement

My anthropological interest in Newfoundland has primarily centred on its outport communities, small fishing settlements that once encircled the island, but are now becoming increasingly rare. The reason for their erasure is a series of resettlement programs put into place when the island joined Canada in 1949. Designed by Newfoundland's first premier Joey Smallwood, the system was intended to centralize remote populations into so-called "growth centres" by offering fairly sizable financial incentives for residents to simply walk away from their communities. The idea behind this set of policies was to bring the province into the modern era with a move away from salt fish production to fresh-frozen based fisheries. Along with this perceived modernization came amenities such as sewer, electricity, roads, access to better education, and improved medical services (Mayda 2004).

From a political and economic perspective the policy made sense, but from a socio-cultural angle, it was seen as the forced removal of a core element of the island's cultural makeup. The sense of cultural evaporation fully bloomed from 1965–1970, when 16,000 people from 119 outports were resettled. After 1970, only a handful of settlements voted (90% of residents needed to vote to resettle to receive the buyout) to abandon their communities (Mayda 2004). Still to this day, a number of outports cling to this way of life, periodically tabling a vote for resettlement, sometimes tipping over, other times holding on for a few more years, even decades. Grand Bruit was on the precipice of being resettled, and I thought it would be interesting and useful to spend the summer with the 15 remaining inhabitants to catch a glimpse of a place in the process of becoming a ghost town (for a more in-depth outline of the various resettlement programs in Newfoundland, see Parzival Copes' 1972 report *The Resettlement of Fishing Communities in Newfoundland* for the Canadian Council on Rural Development).

### Grand Bruit

From an old kind of fog, a tiny inlet opens onto a settlement that looks deceptively populated. As the ferry slowly floats into the harbour, I see a cluster of people and ATVs, dimly lit by the solitary wharf light. Every one of the village's 15 remaining occupants is there, waiting to see what deliveries they might have, and if anyone new is on the ferry. As the boat slides next to the pilings, I look up at the rainy faces staring down at me, wondering who I was and why I was there. After climbing up onto the wharf, I hear a voice from the back of the crowd ask if I am Justin. She tells me to follow her as she speeds off into the late-afternoon dampness on her ATV. I follow her with my eyes, worried about tracking her route. Later I realized how pointless my attempt had been: there is only one road in Grand Bruit, and it is a sidewalk. No cars have ever been here.

After having spent much of the previous year conducting dissertation research on the ghost towns of the High Plains (see Chapter 8), I'd decided to turn my attention to another island, and to a place teetering on possible disintegration. My intentions were not voyeuristic; rather, they were framed around the idea of a moment in time, in this particular time, a time out of mainland time, filled with the potential for dissolution. I wanted to bear witness to the final layers accumulating and sifting themselves together. I wanted to try to understand the world of these last people riding out the final wave along this forgotten coast.

Everyone in the village referred to the place that I was staying as the Old House. It had been the home of the parents-in-law of the woman I'd rented it from. They had both died quite recently, and the home had been left unchanged. Its wonky linoleum and careful curtains became a readymade museum of outport life, calling out the various epochs and events. The wallpaper called out loudly, its deep layers forming direct strata of time in place, chronotopic decor that bubbled and rerouted itself, stretching around heavily painted window frames. Scratches and tears opened a portal to another universe of meaning, each glimmering a gentle battle of colour against the cold grey of the North Atlantic. Thin slices of island time caught between paper and paste.

Grand Bruit is divided by a large waterfall (Grand Bruit is French for "great noise") spanned by a small footbridge. The settlement can be walked from end to end in about 15 minutes. Early in the summer, I developed a route that I walked a few times each day, drawing an everyday line to familiarize myself with the people, places, and things of the village. At one end of my line was a warped wooden causeway connecting a small rock outcropping with a single house on it to the centre of town. I was told that a wealthy American had purchased the house and was allowing it to fall to ruin. Soon the causeway would crumble, untying the micro-island. I wondered after the house and its absentee owner, building my own tiny narratives from my bedroom window in the Old House. A mysterious island hovering at the edge of my days, but for the people of Grand Bruit, it hovered at the limits of years. More strange temporal equivalencies.

As humans, we assemble and reassemble our lives and the lives of others, constantly writing and rewriting narratives about the hidden meaning of our world. Islands (in all of their varied forms) are a special focus of this imagination, as is the peculiar condition of abandonment. Desert(ed) islands and the haunted house loom large in our collect(ed)ive dreamworlds. We inscribe our wonderment and fantasy on these places, carefully considering our intersections and alignments with them. As anthropologists, our vocation is precisely this kind of speculation. We are always attempting to anchor this imagining and wondering in a mode of thinking that asks and addresses the core of our questions to place them into specific cultural contexts.

Still, the particular provenance of this lost place remained a mystery to me during my time in Grand Bruit, but perhaps more important than apprehending these details was the answer to my questioning; not my questions themselves, but the process and outcome of the practice of inquiry. In many ways I was more interested in understanding why I wanted to know, and by knowing, how I hoped to encounter a form of knowing. An absent dreamer opens the door to my dreamworld.

In Grand Bruit, I stumbled into a kind of phenomenology of islands. I began to experience a sense of place directly as a lived reality. The tactile and sensory elements of the village gained new resonance, and the meaning of remoteness and distance felt sharper and more whole. Grand Bruit's islandness emerged less from its location on an island, but more from its isolation. Returning to the notion of layers that I discussed earlier (see Chapter 3), a phenomenological approach to being in this place produced a clearer sense of the multiple and intertwined genres of island. Newfoundland-as-island formed and informed Grand Bruit-as-island, the anthropologist-as-island was thrust into conversation with the Self-as-island. A good deal of these revelations emerged from the simple practice of walking around the village and its still hinterlands. Echoing de Certeau's (1984) concepts of *strategy* (codified, bureaucratic elements of our everyday lives) and *tactic* (the improvisational response to defined strategies), I see my movements in Grand Bruit as a set of tactics that resist the strategies imposed by cartographers, governmental agencies and the general by-products of enculturation. A phenomenological engagement with islands offers access to the multiple layers of islandness. And as an ethnographer, it is my duty to translate these varied and variable experiences into forms of knowledge that make this view portable and relatable. For de Certeau, the bird's-eye view (strategies) cannot compare to the ground-level direct engagement with all of its associated messiness and improvisation.

## The Practice of Everyday Islands

Returning more directly to de Certeau (1984), I suggest that islands and life on islands form a particular kind of accumulation of tactics, making them, following Levi-Strauss (1966), *good to think with*. Here, considering the previously outlined notion of layered place on islands, a potential for the hybridization of de Certeau's concepts of tactics and stratified place, or what he has described as the "piling up of heterogeneous places" (201) becomes possible. And with this synthesis at hand, there is potential to understand and engage with island people, places, and things as distinct forms of collage that incorporate various endemic tactics and strata. This particular outlook can be called *island knowing*. From my perspective, this tool is deeply enculturated in the islander, but something that has to be developed and continually refined in those of us *from away*. And as previously mentioned, I certainly acknowledge that we cannot simply distil real life experiences into convenient anthropological metaphors, but I think that there is space to consider the ideological island as a location for critical reflection, and one that exists in a productive parallel dialogue with the geographic island.

This layered application of an anthropological perspective can also be understood as a particular mode of approach, a tactic for the manipulation of a set of observed phenomena in an attempt to form an ethnographic engagement with islands. In this phenomenological experience of islands, it is important to ask what we (the Ethnographers) *do* with these wild and varied tendrils of meaning. What do we fabricate, collate, and narrate with our careful attentions and awarenesses? We have the luxury of being the writers tasked with the translation of the magic of the everyday life. In some ways this notion opens the possibility for the emergence of the practice of islands as a bounded set of critical lenses. It asks how the *etic* (outsider) grammar of anthropology can assemble the stratified narratives of the place.

## The Sound of Outports: On Islandic Audio Ethnography

Like Jupiter and its moons, some islands have satellites: dreamy, disparate, and rocky. Fogo Island, off Newfoundland's northeast coast, is certainly a moon-island. Going there for the first time, I'd been asked to participate in a month-long workshop to develop concepts and approaches for a new luxury hotel being built on the island. A friend of mine working on the project, originally from Newfoundland, had encouraged me to apply as someone with a certain anthropological perspective and a familiarity with outport culture. Soon thereafter, I found myself en route to Fogo via Gander, Newfoundland and then on the ferry where I would begin the next phase of my relationship with Newfoundland and islands.

Touching down in Gander feels like time has slipped its harness, like entering a could-have-been transatlantic future. The Gander International Airport has become a strange anachronism, a hauntological dreamworld filled with mid-century hopefulness and visions of prosperity. Now, the international departures area of the building is locked away, and the domestic section of the terminal appears more like a failing Midwestern mall, right down to the saturated browns and oranges that bleed carpet into furniture into chipped tile outside the bathrooms.

Like everything on an island, the soundscapes are bounded. Often uninterrupted and feral, enclosed by ocean currents, they have difficulty swimming far from shore. During the development of the Fogo Island Inn interiors, part of my work was to collect and assemble the local aural landscape into ambient sculptures to be used as a means of connecting the external world of the island with the Inn's inside spaces.

The challenge of reflecting the audio world of an island forced me to consider how the phenomenological approach to an ethnography of islands might be translated into sound rather than writing or images. What do islands sound like? Beyond the crashing waves and gulls, what sonic textures emerge from *being there*? Soon I realized that each sound is an island, part of an archipelago of noises and notes, connected by strands of experience. In collecting and assembling the recordings, I found myself once again performing a Levi-Straussian (1966) *bricolage*, assembling the meanings out of readymade elements, making a collage to draw lines (Ingold 2007) between aural islands. Then this idea draws me back to considering how to take the fragments and nodes of islands and reform them into new kinds of awareness. How might I be able to reassemble an island? My intent is not to rewrite or overwrite, but rather to create another potential island, a new order of accumulations.

# 5 Iceland I

## New Old Dreamworlds

### Beginning Iceland

I first went to Iceland in 2006 as a new graduate student, excited and wide-eyed in the nightly summer sun. I'd long been fascinated by this place as an abstraction, as a sort of essential island at the periphery of my imagination. I'd caught glimpses of Reykjavik in the Sugarcubes video for "Birthday", and I'd seen Baltasar Kormákur's 2000 film *101 Reykjavik*, which was enough to set the place in my dreaming mind. Before Iceland's 2008 financial collapse, it wasn't the overblown tourist destination that one is likely to encounter now; at that point it was relatively rare to know someone who had travelled there. For me, visiting Iceland seemed as likely as going to Easter Island or Madagascar. Another kind of phantom island, more of a word than a place.

A few years earlier, a flight I was on got diverted to Keflavik, and when we landed, I begged the flight attendant to let me off to no avail. I'm not quite sure what I thought I'd gain by wandering around the airport for a few minutes. I suppose the sense that I'd been there? Crestfallen, I stared out across the tarmac, the whispers of the surrounding lava fields flickering on the misted window. I wondered if this was as close as I'd ever get to Iceland.

Luckily for me, a former professor of mine had gotten a grant to do some research on Icelandic arts communities and their connection to landscape, and part of the budget included a student assistant. She asked me to join her in Reykjavik, and soon I was walking off the plane in the Keflavik airport. And while this was not my first encounter with islands and anthropology, it was certainly one of the most formative. Over the next three weeks, I steeped myself in the dream I'd created for myself. I let the waves of *there* wash over me, picking out bits and pieces to bring back *here*. I felt like I'd fallen into a "Birthday" and *101 Reykjavik* vortex. Arriving on the island of one's imagination turns everything resonant; everything flows and glimmers. Some islands are like the long-lost cousins that we never knew we had.

DOI: 10.4324/9781003297581-5

## Wandering (in) 101

Drifting through the 2006 streets of downtown Reykjavik was quite different from getting swept up in the 2016 crowds as they pushed their way toward Hall-grímskirkja, the phallological museum, or the hot dog stand. In a pre-collapse Reykjavik, the city felt calm, pleasantly distant, and open. I felt that the locals had a gentle disregard for visitors and that Reykjavik's performance of itself had not yet ballooned into its current state. I remember stumbling across an abandoned office building at what was then the outer reaches of the city centre, a place reclaimed by artists and musicians. Vaguely European techno gurgled out of its gap-toothed window holes and its facade dripped in vaguely New Yorkian graffiti. The idea of such dereliction in today's Reykjavik seems unimaginable. Any wrecking balls have given way to cranes, and all is hotels.

During this first trip to Iceland, I did a lot of unplanned wandering. Sifting through residential neighbourhoods and the quieted waterfront as the sun barely set, only to resurface a couple hours later. With a surplus of time and a deficit of money, drifting became my everyday practice. Drifting like water. Bumping into, floating over, and beaching myself on islands on an island, making tiny city aquapelagos where I am both island and water.

The coffee shop *Moka* is the last holdout of an older Reykjavik, a not-too-distant past of 101. It is an island amidst the sea of tourists and glittery fish-and-chips shops backed by places selling volcanic cosmetics. And as everywhere, islands rise up out of their water to quietly make themselves known. This little wood-lined cafe was not such an anomaly in 2006, but now it has become much more of an island. And just like the anthropologists before me, I can see it through my positional lenses and wonder about its resonance and consequence in the world. Nowadays it's one of the few places where a visitor might hear more Icelandic than English.

In these early years of my Icelandic attentions there were deserted islands, even in 101. Worn-out houses with peeling and stuttered paint, forgotten alleys and pass-throughs where graffiti flourished in the absence of the Instagram gaze. And as anthropologists we often hold our places in a special light, our islands. These are the places of our dreams. We must remember that no island belongs to its anthropologist, yet we can harbour layers that float above, along, and beneath. Older Reykjavik is one of my layers, calmly bobbing along above the heads of tourists and locals alike. No better or worse, just one layer among many.

## The Mountains and Silences of the Westfjords

Looking back at my first encounter with the remote northwest of Iceland, my memories are more of a bird's nest of impressions and resonances than anything concrete, a collection of overlapping energies. Walking in the hills and moss-scapes north of Hólmavík, I felt for the first time an old power that would come to haunt me, something that curled its blood into my blood, a sense of place that reminds us of things we can't remember. And once again I was reminded of the resonance of islands, of their echo-chambered clarity. Still, I was careful not to anchor these sensations in the supernatural. It was something approaching the extra-human, something just outside and unnameable, a sense that has followed me and reappeared in many places and times.

As a brief detour from my work in Reykjavik that first summer, I had decided to explore the almost-island of the Westfjords region (the peninsula is connected to the rest of Iceland by a 7 km wide isthmus). I had taken a series of buses of decreasing size to reach Hólmavík, where I had rented a room on a sheep farm a few kilometres outside of town. I spent the next few days wandering the valleys and bays around the village, trying to absorb the weird energy of this place, trying to capture little bits of it to ferment in my slowly unfurling dreamworld of this place. And in all my anthropological being, I was working to *be there*. I was interrogating the landscape to begin to understand something about its distinct islandness.

The mountains in the Westfjords are haunted pyramids. They loom and hold, casting dark invisible shadows on the landscape. Tombs and mausoleums for undead time, something locked away with narrow windows to look longingly at the new world. Long white fingers wrap themselves over the peaks and they stare down through my window in the misted evening. Humans are a scratch on the timescale of these giants, and like geological megafauna they echo back to all times before. Sitting among the pinpoint blue flowers with these considerations reverberating in my head, I wondered how this could be anthropology, and also how it could not. In the crystal moments that islands impart, the world gets its tender luminosity.

Nowadays, after over 15 years of coming to this place, the mountains still ripple and vibrate; they lunge out of the fog as I round a muddy corner near Bíldudalur. They surge through the resting clouds like an imaginary whale breaching, spewing their existence into the world. They are all the remains of a vanished civilization, a city of nobody. Not the mountains themselves, but the vapor of their meaning. A narcotic purity of place that generates and follows envelopes of being. The village disappears below the ridge and all I see is the far shore. Snowy silence and lit-up rock across the time-out-of time fjord.

A quiet without people is silence, a pure quiet where human inscription is afraid to walk. It is a rare unsound that can be only non-human. Pure quiet is the absence of culture, to sit with itself in lengths of unmeasured time. The human remnant, revenant, fades to leave the soundless sound of a pair of snow geese landing their white bodies on the mossy rubble where valley becomes mountain.

2021, late autumn, Reykjarfjordur: the water calms itself in the fjords; not like glass, more like cellophane stretched tight, still crinkled in places. Seal heads and seabirds punctuate the flatness with round eyes and wings. Floating on my back in the warm water of a geothermal pool, I feel like an island. Perhaps this moment is participant-observation? Perhaps it is more ethnographic method acting. Am I attempting to dissolve into this place in the way that I had hoped to embody the ghosts of the Faroes? To begin to know islands, maybe we need to become islands. Or maybe we are always-already our own private islands.

## Iceland through My Students

I have been taking small groups of students (no more than eight at a time) to Iceland as part of a field course in ethnography for over 8 years now. What started as an experiment in teaching island ethnography has become one of the primary reasons I have such an abiding affection for Iceland's people, places, and things. In the context of this course, I have come to realize the significance of the practice of teaching as a mode of knowing, and I have developed a much more nuanced sense of my work as a writer and ethnographer through this endeavour.

In some ways, I have become an ethnographer of my students in Iceland, attempting to unpack their micro-cultures and untangle their unique insights. I place their views as islands in an archipelago of understanding. Each day we are there, each year we return, I gain a new viewpoint, a new way of being and seeing. And as I teach about islands, I learn about islands, absorbing the novel visions and unencumbered insights of my students.

Since we spend the bulk of our time in the Westfjords, threading our way through toothsome fjords, we are never far from the ocean during these trips, a reality that constantly reminds us of our boundedness. This isolation grounds our conversations; it forms and informs our internal maps of this island. If, as I suggested earlier, islands are *good to think with*, perhaps they are also *good to teach with*. And perhaps they are not only good for teaching our students, but also for teaching ourselves.

## Layered Dreamworlds: Veils

The veil is thin in Iceland, much thinner than in other places I have visited. This notion of the veil, a kind of half-transparency floating over everyday lived realities, is not necessarily supernatural. The veil is the ether that hides Benjamin's (2021) profane illuminations, the pinprick moments of realization when a parallel way of being and seeing makes itself known. Realities collide, converge, and converse; they overlap and inform. This notion of layers has been discussed elsewhere in this book, but I feel that it is important to revisit it in the context of Iceland because it is here that my sense of these layers is most resonant and present.

I have sometimes claimed to be a cultural geographer, examining the intersections and connections between humans and landscapes. Of course, in this engagement, my attentions turn to islands and their unique propensity for layering themselves and for allowing those layers to breathe, move, and define themselves. A careful and purposeful notice allows the anthropologist of layers to see the spaces between, to attune themselves to the flutter of the veils that lay densely on islands, and especially in Iceland.

Yet the simple explanation that 54% of Icelanders believe in elves is a blunt touristic fact. There is something that haunts the landscapes here, a wild and spectral sense of place. There is some unnameable energy in the everyday lifeworld of this place. In lifting the veil, we see the hidden worlds and we can begin a taxonomy of veils and everything that we find behind them.

## Flatey

Flatey is one of a multitude of islands in the constellation of Breiðafjörður in Western Iceland. Another island of an island, Flatey was once a cultural and economic centrepoint for this part of the country. Now only a handful of people live here year-round, and summer people haunt the haunted houses of their forebears.

Flatey is another example of island relativity. Iceland becomes the mainland, and this place becomes the satellite. It is, as its name suggests, quite flat with a rather unremarkable topography. A cluster of painted houses, including a lovely hotel, rest at the end of a dirt track leading up from the ferry dock, and seabirds swirl overhead. Sheep rest watchfully from tufts of grass, and everything appears truly bucolic. No longer the artistic and political hub it once was, the island is now a curiosity for tourists and refuge for summer people from the city. It is an island set adrift, now tethered only by its geography, its social and cultural gravity diminished.

Last I heard there were only five people living on the island through the winter. Like astronauts on a space station, holding the line of inhabitation at the outer realms, pushing back against the nothing.

11PM light streams in through the narrow window in the hotel's basement pub, washing the worn floorboards with golden dreams of summer. I talk with the art student behind the bar who is working there during the tourist season. Everything is calm and warm, the evening opens up into forever over the bay, and the ferry ride out tomorrow is both distant and close. Lying in my bed upstairs later that night, a new spectre creeps in under the door. Winter on the dark side of the loop, and life on the moon feels a little less perfect.

## Roads

Black-grey tendrils warp and weft themselves across the Icelandic land-scape, the feral cousins of the paved routes and Grandmother Ring Road that slip through mountain passes and skirt unbound cliffs. Hints and haunt-ings of not-so-long-ago that call out to a rough-cut existence on this island, the old paths of before. These dirt tracks become a kind of portal to another way of knowing landscapes. They are silent siblings that speak to a pre-touristic epoch.

I stop my 4x4 and step from the warm glow of its interior, out into the day-for-night wind. Out at the edge of 622, a one-lane road built in 1973 by a local farmer using only a small bulldozer, I am once again an island. Cut free from the vehicle that brought me here, floating at the boundary of here and there. More silence in heavy syrup, washing and clinging to me in the shadow of an abandoned lighthouse. A little further down the road, a deserted farm with tattered plastic flapping in its empty windows. A sheep skeleton bleaching out in a shallow ditch. The mountain roads usher in remoteness, guiding hungry ghosts to find their solitude.

In many ways, these roads are islands unto themselves, drifting at the gravel periphery, bound by impenetrable depths. 622 (known locally as the Dream Road) now leads to a pure kind of nowhere, a platinum nothing. What was once a link to the outside world now becomes remoteness. This road carves off a piece of Iceland and puts it out to sea without inhabitants and with irregular ferry connections. These roads are a portal to another, more islanded Iceland. One that is less integrated and more remote; they are a window into another layer, another kind of island.

## Rings of Hornstrandir

Perhaps islands are objects. Imbued with the same potential for agency, the same ability to circulate and infect. Maybe they are *things* more than they are *places*. Maybe they are people, with emotions and dreams. Maybe we made them, and maybe they made us. Perhaps all of these views are true, or perhaps they are just the magical thinking of an anthropologist in love with islands.

Iceland's Hornstrandir region has been abandoned since 1952. Documen-tation of the human history of this region has been limited, and even less is available in translation. Aside from a few tourism websites and the odd blog post, there is little information. It is almost as if the region has been abandoned not only by people, but also by historiography. Still, it seems as though almost every Icelander I talk to has some form of familial or cultural connection to the place. It hovers at the edge of a nostalgia, haunting the national imagination as a kind of pre-industrial fantasy, a hardscrabble utopia at the limits of the island.

More un-night seeps down into the grey valley from Hornstrandir's interior, in glassy rivulets and imaginary windchimes that I can almost hear on the waves below. The cool energy clicks and pops through mossy undercurrents, rising and falling in crests of sound en route to the Arctic Ocean. Further down the valley, my tent glows orange amongst the smothered foundations of ancient farmsteads. Tufts of errant fox fur cling and flutter along paths used long before we arrived here. Like W.G. Sebald (1999) on his walk through Suffolk, I came to this remote claw of the Westfjords to trace a line across the northern coast on foot. In a hybrid of method-acting and ethnography, I wanted to haunt the haunted to perhaps open a miniature window into the *genius loci* of this place.

One step deeper than the remoteness of the F-roads, these pathways scar the more remote of the already remote. Much as I had attempted to embody, in some small way, the last days of a remote village in the Faroes (see Chapter 3), here I had hoped to ingest the essence of this abandoned peninsula that just barely evades the Arctic Circle. I wanted to retrace the lines made by a thousand years of people, pathways between isolated farms in isolated bays, an archipelago of resistance, people whose now-sunken homes once hacked dreamworlds into the face of this savage hinterland.

Like other islands, Hornstrandir is only reachable by boat (or a walk of several days), and then, effectively only for a couple months during the summer. For most of its time, it is dark and cold, orbiting the relative warmth and humanness of the populated places of Iceland. It is an abandoned archipelago of once-upon-a-time farms and weary footpaths that make themselves into rings, lopsidedly encircling the planet of people, fragments of exploded dreams caught in the gravity of remembrance.

# 6 Iceland II

## Come-from-Away

### Authenticity and Being There

1.5 million tourists are predicted to visit Iceland in 2022. For an island with just under 400,000 residents, this number reflects a distinct form of cultural change playing out in the country since its 2008 economic crash. In a (quite successful) attempt at recovery, the nation doubled down on its potential as a tourist destination by centring its natural beauty and northern exoticism in various marketing campaigns and promotions. In many ways, this plan worked too well. Between 2010 and 2018, visitors increased from 500,000 to over 2 million per year (Sæþórsdóttir, Hall, and Wendt 2020), leading to a condition that has often been described as "overtourism", wherein the ecological, economic, and socio-cultural "carrying capacity" of a place has become strained and unsustainable (see Geoffrey 2020). This issue becomes particularly resonant and palpable in an island setting because as peak carrying capacity of visitors approaches, there is no space (ideological or geographic) to expand, and the island begins to sink under its own touristic "success". Iceland is increasingly oversaturated, overwritten, and overwrought. And at the core of this touristic fetishization of Iceland is the quest for authenticity and a genuine sense of *being there*, a drive to reproduce some form of touristic truth at the imagined edge of the world (see Kristín 2019).

Iceland becomes a gleaming beacon in the eyes of the would-be adventurer, a collection of lava fields, volcanoes, and geysers endlessly looped on social media feeds. This view of the island rests firmly in the idea of nature-based tourism, a focus that acts as a kind of counterpoint to cultural tourism in places such as Paris or New York, where cultural forms of touristic encounters dominate. This attention to the unspoiled natural landscapes finds a home in the vision of an epic sojourn to the North, to the limits of the European periphery where strange and beautiful environments prevail (Oslund 2011). For many, it is the true "idea of North" as described in Glenn Gould's 1967 radio documentary by the same name; it is approaching the unapproachable,

DOI: 10.4324/9781003297581-6

the production of an ontology of remoteness. It is a distillation found carefully inscribed in the safety videos and in-flight engagements on Icelandair that remind tourists of how and why they should experience the country, validating their rugged dreamworlds. And again, even in this liminal existence of the flight, visitors are conditioned and enculturated, reaffirming their belief in an authentic experience. In this way, the plane is a kind of ferry (see Chapter 1), offering a gentle liminality en route to a similarly gentle ideological wilderness: safe intrepidness, authentic remoteness without the discomfort. And as these pilgrims of *away* thumb through their guidebooks and itineraries, all of their Instagram imaginaries take shape as they sleep their way across the North Atlantic. They move between nation states and states of being: from mainland to island, from local to visitor, from observer to actor.

From an anthropological perspective, it is important to consider the touristic carrying capacity of Iceland from both an ecological and ideological angle, especially in light of the production and consumption of authenticity, one of the primary commodities embedded in tourism on the island. I have seen this change first-hand over the last 15 years as I walk the once-calm streets of Reykjavik's downtown. Each year this place becomes more of a performance of itself, quickly forming into a simulacrum, a copy without an original (Baudrillard 1994). As rental cars dodge selfie-snappers and urban trekkers armoured in Gore-Tex, puffin paraphernalia and plastic Viking helmets burst from every second doorway along Laugavegur, the main tourist artery of the city. Now it seems as though every building is a hotel or vacation rental; every street corner is the pick-up and drop-off for a guided tour. Cranes and cement mixers hum and growl behind slapdash plywood walls. And this rolling colossus of enterprise bleeds out across the island as roads get paved, tunnels get built, and the countryside becomes another series of nodes in the interior archipelago of Iceland.

The Ring Road draws tourists toward the sparse majesty of an Iceland of pristine nature and the solitude of a preternatural world. Yet as Gísli Pálsson (2013) claims, a truly unpeopled place is not what these nature-seekers hope to find. They want the hints of humanity and otherness embedded in the landscape; they want tumbledown farmhouses and signs of Iceland's struggles to help form the objects of their touristic desire (Lury 2002). They want a glimmer of mystical Iceland with its elves and trolls, its otherworldly mythoscape that becomes a perfect commodity to be exchanged for validation of their authentic experience when they return *home*. And again, these elements emerge from a uniquely islanded place; their singularity and authenticity are the direct result of their boundedness. Iceland's islandness renders it undiluted and untainted, authentically Iceland.

This dreamt-of Iceland forms another layer on the island, one that is continually rolled out by tourists and those seeking tourist money, and it grows deeper and wider every year. These collections and layers of experience take shape as a unique kind of cultural capital, a form of capital that produces (and exchanges in) rarity and intrepidness. Initially outlined by Bourdieu in his well-known "Forms of Capital" (1986) article, cultural capital can be defined as the embodied knowledge generated through enculturation, education, and immersion in certain kinds of cultural milieus. Travel—and more specifically, the kinds of travel and chosen destinations—is a key site of production for cultural capital. The value of this capital is largely dependent on what might be viewed as the *sphere of exchange* (Bohannan 1959) in which it exists. Amongst seekers of remoteness, adventure, and unspoiled nature, Iceland holds a high potential for producing cultural capital and generating perceptions of authenticity for tourists. Having *been there* offers the tourist the *aura* (Benjamin 2008) of authenticity; they are delightfully haunted by their true engagement with remoteness (Hastrup 1998). Nowadays, that value has decreased due to a rise in what might be called *touristic inflation* and the relative evaporation of the rarity of visiting Iceland. When greater numbers of people can travel to these places, the value of their cultural capital plunges and they return to the sea of ordinary destinations. As Iceland gets increasingly incorporated into the global system of tourist destinations, the cultural capital of these authentically rare experiences starts to lose its value within the spheres of exchange in which it once figured so prominently. The diminishing difficulty of getting to Iceland and the increasing commonality of this experience begin to erode the value of this kind of capital, pushing visitors further out to the hinterlands of Iceland (the once remote and relatively untouristed Westfjords region was just named Lonely Planet's Best of Travel for 2022) and on to other more "valuable" destinations such as the Faroe Islands (see Chapter 3) or Greenland. As experiences of remoteness become more commonplace, they become subject to a devaluation and the cultural capital begins to fade. A large part of Iceland's touristic appeal has been its northern remoteness and island singularity, and now, as a result of overtourism and global interconnectivity, its authentic value is poised for a sharp decline in the near future.

## Ethnography and Tourists

In many ways islands are perfect touristic destinations. They are contained and knowable, yet distant and exotic. They are inarguable peripheries to the everyday drudgery of mainland existence. An ethnographic attention to tourists, and specifically island tourists, asks another important question about authenticity, but rather than directly interrogating the authenticity of the touristic experience, it is important to consider the age-old dilemma of the authentic subject of anthropology. I believe that all people, places, and things are productive sites for anthropological inquiry, and as such, tourists become a perfectly viable ethnographic subject. They (and if I'm being autoethnographic here, I might say "*we*") are a mass, an ideological wave cresting and breaking at arrival and departure gates. They are here as they are everywhere. They are shifting and disjunctive, forming unique kinds of fictions and frictions (Tsing 2005).

I cannot divorce myself from this group. I am also a tourist, everywhere and every day. In many ways anthropologists are professional tourists, albeit with specific goals and practices in tow. Like tourists, we are interested in how other people, places, and things undertake their everyday lives. In this way, I feel that my work as an anthropologist is a kind of intellectual harnessing of my touristic soul. I often struggle with a sense of disavowal, pushing against my impulse to separate myself from other tourists by employing (or hiding behind) my anthropological lens, as if it offers some sort of escape hatch from being swept up in the throngs of (other) tourists.

My approach to the duality of the anthropologist-as-tourist, is to adopt a dialogic perspective wherein I can be both tourist *and* anthropologist. To cast out the tourist is to deceive myself and my readers. I am a tourist of tourism, and an anthropologist of anthropology, a visitor and student of islands. Iceland is the perfect setting for considering these intersections because it presents a variety of forms of both anthropological and touristic encounters. Iceland and its tourists are both an existential and professional crisis for me. They form a microcosm of endless anthropological potential.

## Touristic Oriented Ontology

Tourists in Iceland form their own layer, a previously absent ethnoscape (Appadurai 1990) that overwrites and underwrites the island. New values-capes explode out of nothingness to create unknowable yet wholly pervasive ways of being. Locals perform their culture for visitors in a slightly detuned utopia, helping to strengthen, expand, and deepen this layer. In considering how tourism (re)makes the cultural fabric of an island, it is important to ask how and why it might be useful to consider the tourist as the author and audience of a new genre of dreamscape, one that shapes and reshapes the real and imagined spaces and places of Iceland to create a kind of heterotopia (Foucault 1986), wherein this layer becomes another parallel world existing alongside Iceland's pre-tourist base.

Returning to the notion of layers of meaning in place, Iceland can be understood as a duality, a pair of collections of various layers of meaning. These layers can be described as the local and the global, the touristed and the inhabited, the old and the new. And despite its relative youth, the touristic layer dominates, becoming both thin in historicity and thick in cultural influence. It is pervasive and enveloping, and in some ways nearly every place on the island is in the process of being overwritten as the tourist map covers almost every corner. The map has almost become the territory (Korzybski 1933). And this apparent inevitability generates an island-on-an-island, a parasitic covering that will one day fade into the background.

Reflecting on this duality of place, I am reminded of Arkady and Boris Strugatsky's 1971 science fiction novel, *Roadside Picnic*, in which extraterrestrial visitors arrive on Earth for a brief stay, leaving behind strange castoffs with unknown uses and powers. In thinking about Iceland and tourists through this lens, once Iceland has become depleted of its touristic value, perhaps this layer will dissolve, exposing a base of ideological and material remnants with weird powers and hidden significance. The ruins of occupation lie in place as the layers slowly recede, like the glaciers that carved and gouged the landscape to reveal the natural landscape now coveted by the touristic swarms.

A touristic oriented ontology sees the tourist as author and executioner, and they/we write worlds into being and subsume place with their/our presence. The tourist brings new modes of knowledge production, new practices of the everyday, and new layers of meaning. To see the tourist as an ethnographic site renders them as both lens and mirror for the landscapes, ethnoscapes, and ideoscapes (Appadurai 1990) that they produce and occupy. At some point, any touristic place begins to equalize, and hosts and guests coexist in their respective layers, sometimes weaving in and out of each other's worlds. For Iceland this moment is still a future, and the tourist wave has yet to crest. And it is in this moment of pre-collapse that the potential for ethnographic insight bubbles up.

## The Anthropologist-as-Tourist

And I am still, as all anthropologists, a tourist among tourists. Despite my anthropological intentions, I am always-already a tourist. I float amidst the flotsam and jetsam of other tourists, and among the Icelanders and ethnographers. I will be the first to acknowledge my non-localness (or maybe someone at the airport gate has already silently called me out) and I recognize my place as a visitor. As outlined here, I believe that ethnographers are always, somewhat unavoidably, tourists. And while our motives and practices take varied shapes, we all want something from elsewhere, some object captured and held, a thing to prove our presence and authenticate our experience (Lury 2002). So then where do these insights come to rest, and what usefulness do they have in helping to make sense of the everyday lives of anthropologists and tourists? How are the Anthropologist and Tourist cousins and travel companions? Perhaps we are a two-headed monster, never quite certain who is in charge.

# 7 Phantom Islands

## Shorelines without Islands

### Uranium City

My grandparents lived in Saskatchewan their entire lives. Sitting at their kitchen table on my way up north to Uranium City, my grandfather asked me to point it out on the road map I had unfurled on top of the remnants of our dinner. The map ended less than two-thirds up the province's rectangular form, rendering the northern third of the province a tiny inset with a few faint lines for Uranium City's Highway 962, and further west, the mysterious 999 around Camsell Portage. Both of these roads are transportation islands bounded by forest and rock, unconnected from their southern kin. And even on this map, Uranium City is a boxed island in two dimensions, cut off from the rest of Saskatchewan.

My grandfather had a vague recollection of the name, but nothing more. A town built in the far northern corner of the province, clinging to the shores of Lake Athabasca, Saskatchewan's second-most northerly settlement (after Camsell Portage) is no longer the mining boomtown it was in the late 1970s. By 1981, the settlement had over 2,500 residents, and people were moving in from around the province and the country. This sense of prosperity was short-lived and with the shutdown of the mines in 1982, the population fell drastically to only a couple hundred. The closure of the regional hospital in 2006 brought the number to just above 100, and the 2016 census indicates only 73 inhabitants remaining. A hauntological city in the woods, built for thousands, now occupied by far more ghosts than human souls. Islands themselves, disconnected at the far reaches of a different kind of ocean.

At the outer edges of "town" are an uncanny haunting of hastily emptied bungalows and never-occupied apartment blocks with open mouths that gape into the trembling subarctic forests. Cul-de-sacs devoured by taiga entropy and felled streetlamps enveloped in moss form a never-to-be suburb. Here, I am an archeo-ethnographer, sifting through the island's rubble (Gordillo 2014), interrogating the becoming-ruins as waves of leaves and

DOI: 10.4324/9781003297581-7

branches swallow old culture. Like Malinowski, my patron saint of island anthropology, I sit with my subjects, looking off across the horizon of surging difference. Even now, as these wayward suburbs are cut loose from the centre, they whip and whisper at night in the language of torn plastic and broken wires. They become outer islands, deeper, more distant rings. They are colder and quieter with every passing year. And they are lines and layers and explosions. Connecting, accumulating, and illuminating. Islands are everywhere, if only we attune ourselves to their signals and noises.

It is not possible to drive to Uranium City, at least in the summer. In the winter, ice roads connect the last remaining residents with the rest of the province. But most people fly in and out. And in this way, Uranium City is once again chopped out of the province, another kind of outport (see Chapter 4), arguably more isolated. Part of what makes places like Uranium City into islands is the ideological rift that often surrounds them, a sea of forgetting. Like waterbound islands, these locations are islanded by their relative invisibility.

The wooden floor in the gym inside Candu High School surges and swells, buckled by rainwater. The building was state-of-the-art in 1982 when it was built, but the school was abandoned like the rest of Uranium City only a few years later. These waves of floorboards, rising and falling in a perpetual freeze, remind me of other kinds of waves crashing against other kinds of shores. There are waves and islands and shorelines that call themselves by different names, and we see them in different lights, but in so many ways they are the same. Uranium City is almost a phantom island now, almost an island of phantoms.

A large part of how and why I came to Uranium City was to document the last remaining lives, to catalog and concretize their narratives, to draw one thin line back to the mainland. I sat on the porch in the forever day of a northern summer, chatting to a family about the struggles and magic of living in a place like this. The two teenage children told me how they missed fresh (as opposed to powdered) milk since it was too expensive to bring up to Uranium City regularly. They also spoke of the natural beauty and solitude that surrounded them. These small points in a much longer conversation resonated with me because they reminded me again of the common threads that tie islands together. These everyday Mainland realities are extraordinary for islanders, both the struggles and the magic.

### Nolalu, Ontario, Canada

Following the line of thinking *with* islands, in some ways I also grew up on an island. And like many of the places I describe in this book, my island was not bound by water. My island was just under a hundred acres of wilderness about an hour west of Thunder Bay, Ontario. My parents moved there when my sisters and I were small, hoping to reimagine an oversaturated city life as a forested, off-the-grid dreamworld. Before we had arrived in the early 80s, the property had been a few different sorts of farms, and most recently, in the 70s, a loosely conceived hippie commune whose material remainders provided my sisters and I with endless archaeological encounters amongst the various outbuildings. Things from other eras washed up on the shore, bringing news from older times and places: a set of hand-carved wooden skis left by the Finnish homesteaders, broken-down Volkswagens beached alongside overgrown logging roads, water-stained square photographs of shirtless people we didn't know. These encounters with imagined shorelines and their things-from-another-time were certainly the beginning of my fascination with the *concept* of islands, but of course I was too young to know that then.

This setting was the antecedent to all of my persistent anthropological curiosities. In this remote corner of northern Ontario, my family was not isolated in the ways that many of the other islands I have already discussed are, but we were isolated from the regular flows of everyday life in the 1980s. We drove into the city once a week for supplies, navigating a series of rutted and flooded roads to eventually make our way to the pavement. Getting to high school involved an hour or more bus ride each way, and at night the silence was singular. I remember lying in bed listening to the distant hum of a far-off highway, wondering where people were going, considering the ease of their movement from here to there. That half-dream of wonder still haunts me. In Múli (see Chapter 3), I spent hours staring across the fjord to the village of Viðareiði with its flickering lights and pinpoint car shapes skirting the base of the mountain, again looking out from my isolation, wondering where and why people were connecting themselves to other eleswheres. An island gaze to a relative mainland.

And in this way, all anthropologists have some personal connection to their research projects, these pursuits that emerge from some deeper place, bubbling up to envelope our lifetimes, burrowing in our skin to hold us accountable for our tiny histories. And of course, I am no different. The isolation of my childhood focused my various attentions and attuned my uncharted awarenesses, probing me to ask how culture is formed and informed in and by these environments. To ask the questions about how islands were more than pieces of land surrounded by water, and how an island was both an interior and exterior landscape.

## The Layered Ghost Towns of the High Plains

I spent the better part of two years drifting through the High Plains. During my dissertation research, the ghost towns of southern Saskatchewan, the Dakotas, Wyoming, and New Mexico, along with their associated dirt roads and broken blacktop, became another kind of archipelago. Dried-out (un)waterways to disconnected dreamworlds formed a new version of the High Plains, a new layer of unhinged time and place made out of ghostly points of remembrance.

Phantom islands are cartographic mythologies, often the result of miscalculation or mistaken identity. They are islands that exist only on maps, sometimes as placeholders for empires, other times as simple mistakes (Liesemer 2019). In my thinking about (and with) islands, the ghost towns of the High Plains are phantom islands not because of their fallacy, but because of their tenuous grasp on existence. These places are constantly subject to political and ideological erasure, and in the moment of their eventual disincorporation they appear as unreal as Atlantis. When one of these places ceases to be inhabited, it becomes a phantom island: a name on old maps, a few busted-up houses adrift in a field off the interstate, a weak-kneed road sign with faded green paint flaking off into an unremembered history.

My research in this area examined the cultural causes and outcomes of abandonment, asking questions about the nature of this condition in light of globalization and the rise of the Anthropocene. How were these places becoming increasingly islanded as shifting lifeways made a new kind of world? I was (and still am) interested in trying to understand something, from an ethnographic perspective, about the lifeworlds of those who had remained in these emptied-out towns. I wanted to hear the stories of life on another kind of island, one that was slowly being subsumed by prairies and the surge of changing social and economic patterns. I wanted to know what life looked like in this layer, what sorts of dreamworlds had been cast to the wind. I wanted to see what the everyday looked like in these undersea islands at the bottom of an ancient ocean.

And again, within my anthropological gaze, the layers split themselves out from one another and a new version of this region floated to the surface of my awareness. Multi-lane highways evaporate, as do settlements of more than a hundred people, leaving a layer made out of wheat fields and tumbleweeds with almost abandoned islands connected by almost roads. Coastlines made out of forgotten sidewalks to nowhere and waves of hot wind that replace the ocean's curls.

These places are a genre of island, one not formed by geology, but one made out of unremembering and the dissolution of old ways of being. And as with many of the examples in this book, imagining these places as islands requires an expansion of definitions and limits. Still, this mode of reflection is not a simple thought experiment or quaint academic preponderance; rather, it is a necessary intervention in the ways that we—anthropologists and non-anthropologists alike—conceptualize and imagine these bounded places. For me, the sense of arriving on the Faroe Islands and rolling into Jeffrey City, Wyoming, are so similar that the physical geography of these encounters is truly secondary. The way that isolation and boundedness form and inform these sites is the most relevant and resonant element of their being, at least from an anthropological perspective.

# 8 Hauntological Islands

## The Cultural Imagination of Island Utopias

Instagram coughs up another sparkling island dream. Travel brochures radiate the white sand of perfect emptiness. From our windowless cubicles we stare longingly into the heterotopic vision of palm-lined coasts held endlessly and unflinchingly in our screensavers. The tropical island has become a popular shorthand for utopia, a kind of synecdoche for a better, more elemental way of life, an antidote to the drudgery of everyday life in the landlocked reality of late capitalism. In our rote realities of production and accumulation, the island serves as a beacon, an escape hatch, and a release valve to the condition of contemporary being. They are perfect elsewheres.

As I have discussed throughout this book, islands are good to think *with*, and here they offer a means of synthesizing and placing the notion of earthly paradise. In this way, the tropical island imaginary forms a particular type of heterotopia (Foucault 1986) wherein the image of this place becomes a layer of desire, a compact and efficient *away* that serves as a counterpoint to the ever-present *home*.

In many ways all islands are heterotopic, in that they are composed of a number of layers of meaning and ways of being. They are places apart if we imagine them that way. And just as Foucault describes the mirror as both utopia (a location of virtual perfection) and heterotopia (a location of difference), so too are the images of tropical islands that we conjure (and that are conjured for us). They are utopian because they are unreal, yet they gesture towards a real place. Yet even this seemingly real place is also a simulacra (Baudrillard 1994), another layer, another heterotopia, superimposed on the local. And in this imposition, can we consider the island heterotopia of desire a longing for pre-capitalist existence, where labour has been erased and we are affluent with time (Sahlins 1974)? This vision becomes a heterotopia of time (or what Foucault refers to as *heterochronie*), a look back at

DOI: 10.4324/9781003297581-8

an imagined life of leisure. Foucault calls this sentiment out explicitly when he describes a return to an imagined pre-industrial idyll that becomes a dual heterotopia of endlessly accumulating time (time free of labour) and time-out-of-time in the form of vacation dreamworlds. The tropical island adds and accumulates its heterotopic layers to build an increasingly hyperreal (Baudrillard 1994) pre-capitalist wilderness imaginary. Island resorts with their all-inclusive packages work to erase both labour and money, freeing the vacationers from their mainland dilemmas. And this is how and why we see ourselves over *there*, while we sit *here* and stare into the loving abyss of our screensavers with their unwavering palms and digitized white sand beaches.

In some ways, the concept of the heterotopia serves as the basis for many of the core arguments in this book, in that it addresses the polyvocal nature of islands and their vast and accumulated layers. The notion of an integrated and stratified islandness made up of coexisting heterotopias is the central subject of my inquiries here, and it offers a perfect vehicle for attempting to untangle the lines and layers that lay thickly in place.

## Phantom Islands of the Future

Phantom islands exist now as cartographic ghosts, imagined outposts that served as colonial placeholders, border markers, or wayward mythologies. On one of Yap's Outer Islands, Ulithi (see Chapter 2), I asked a man what he wanted me to tell the rest of the world about his island home. He told me to tell you that he existed, that his island was there, and that it was real.

Apart from the reality that almost nobody outside of Micronesia has heard of this island, this man's message was not about recognition, it was about the great likelihood of his home being swallowed by the ocean. Ulithi and the other Outer Islands sit barely above sea level, and with the inevi-table sea level rise resulting from climate change, this place and its kin are destined to become new phantom islands, rendered in two dimensions and all but imagined. The Anthropocene also rises with its oceans, drowning in itself, pulling worlds down with it as it slips under waves of its own design. Purposeful and perpetual, it sinks.

It will be too late for Ulithi, and before long it will become another casu-alty, a gravesite with no marker, a phantom island on dusty maps and little-visited blogs from another time. And this is a part of my goal with this work, to offer illuminations of these places, to throw out a point of light. I want to nod to the existence and significance of these places. With this attention and awareness, perhaps we can begin to understand the truly interconnected nature of life in the waning moments of the Anthropocene. Maybe we can recognize that we are all islands in an archipelago, and that life on islands and the life of islands impacts everyone. Can we really cast these future phantoms into a sea of forgetting?

## Of Huldufólk and Ghost Paths: Traversing Immanent Planes of Islands

These layers of being—phenomenological heterotopias—are deeper on islands, forming something that I have come to think of as a *stacked ontology of place*. In Iceland and the Faroe Islands, these layered understandings take the form of a parallel universe inhabited by beings known as *huldufólk*, and in Yap they appear as a network of pathways travelled by the spirits of the dead.

Drawing on Deleuze and Guattari's (1988) concepts of *planes of immanence* and *becoming*, I seek, in some minor way, to unwind the knotted world of coexisting ways of seeing and being in an island world. Here, *immanence* is the state of being within, as opposed to *transcendence*, the quality of being elsewhere. As such, the planes of immanence form stacked landscapes of coexistent heterotopias, deep being develops into deep knowing and inward attentions. The spectral-scapes of the Faroes, Iceland, and Yap are the result of a carefully tuned awareness, one that examines the interior life of humans and transposes it onto landscape. In this moment of ideological inhabitation of the island's spiritual geographies, everything is possible and the divisions between lifeways and deathways evaporate. The ghostly overlaps with the corporeal and these landscapes bend and blend into one another. The bounded nature of islands produces a more resonant and immediate sense of immanence, and a stratified understanding of the world destabilizes to allow layers of heterotropic being to influence one another. In these island immanence's mainland determinations are voided and the delineations become elemental relations and intersections, essentially deterritorializing (Deleuze and Guattari 1988) the stacked ontology into pure relationality and formlessness.

It is important, as thinkers *with* islands, to first see and understand the layers and heterotopias before we can begin a practical and productive dissolution of forms, and then to perform the subsequent reterritorialization. To acknowledge the existence of stacked layers is the first phase in conceptualizing the *within*, *upon* and *of* in these places. Here, objects and their being unfurl in fractals of existence, disintegrating in the moments of their formation. The duality of Mainland and Island converge and reassemble, only to untie themselves into more resonant and marked realities. And we just watch it happen, keeping careful notes, whether we know it or not.

Yet what does this attention to island ghosts *do*? Why should we concern our anthropological and everyday selves with these wild whims of feral objects and associations? Because we are interested in the manner and form of their associations, we concern ourselves with the ways that Yapese ghost paths and the hidden people of Iceland and the Faroes push their way into other layers, how they illuminate and dissipate our compartmental relationships to place. And as with everything that we do in the pursuit of anthropological awareness, we seek the strange in the familiar, the here in there, and the exploded view of the everyday. To consider the spectral layers is to reflect on the ways that we often artificially integrate and compress the heterotopias of our worlds. Here, it is also important to remember that the spectral is a key—arguably, *necessary*—component of the everyday, and our attention and awareness to its influence is an important element of understanding the layered worlds we travel and inhabit. As Rahimi reminds us, "the very space of everyday life is so filled with ghosts that nobody can avoid them—in fact, that the very experience of everyday life is built around a process that we call hauntogenic, and whose major by-product is a steady stream of ghosts" (2021:3) This process of *hauntogenesis* is extremely prevalent in island settings as a result of their boundedness. Ghosts produce and reproduce in tighter, more compact realms, both vertically (stacked) and horizontally (shorelined). Islands once again offer a window for these forms of anthropological disgorgement, disentanglement, and dissolving dualities. And in this way, we are the scientists of intersections, the seekers of *within*-ness.

In Yap, Iceland, and the Faroes, the heterotopias of the dead (Yap) and supernatural (Iceland/Faroes) are more immediate and present than elsewhere, pulsing just below the veil. They pass more closely, drawing lines more directly through other forms of lived reality. And in their island worlds they have nowhere to be but *within*. Apparently Icelandic ghosts cannot swim, and when the subjects of their haunting left the island they could not follow, bound to their looped limits. The loops wind themselves around themselves, sedimenting and amplifying, stacking their ontologies.

In Yap, I was told that one of these ghost paths ran behind my house, following the curve of the steep hill across a ridge to another village. These pathways, invisible to the living, drew haunted lines across the island as they intersected and connected, offering another layer, an infrastructure for the dead. In Iceland, I have often heard of the parallel geography of the hidden people, how their cities, trade routes, and landmarks hover just below the human layer. Their world is hinted at in ours. Rock formations and certain boulders serve as gentle protrusions into our reality, reminding us of the thinness of the veil. And once again, if we consider the breadth of the island forms, what other stacked ontologies might emerge? I imagine that there are more ghosts closer to our surface than we suspect.

## Ruins

Islands are easily untied. They can be quickly and quietly cut from global flows and allowed to slowly fade into literal and ideological ruin. Much like islands, ruins are good to think *with*, and their definitions are also fluid and variable. Ruins can be cultural, political, material, and/or temporal. And when islands and ruins intersect and layer their accumulated meanings, new resonances emerge.

Little Fogo Islands was a small outport community about 7 kilometres northeast of Fogo Island (see Chapter 4) where a fishing community flourished until it was eventually resettled in the middle of the last century. And now these rocky points of land float out at the edge of the edge, a curiosity for day-trippers from the Fogo Island Inn and a summer retreat for locals with cabins on the islands.

While working on a design project for the Inn, I had the opportunity to visit these islands. I had been invited along on an early morning trip to witness the opening up of a small house on one of the larger islands in the group that had been unoccupied for well over 40 years. Given my research interests in ghost towns and remote islands, the woman who was buying the house had asked me to come along to document the trip.

As the small wooden boat wound its way through the swells, a few distant globs of land floated on the horizon. As we got closer, houses, a church, and some outbuildings took shape. It was as though the island was forming itself in front of us, emerging out of the ocean, its edges taking on more defined forms. The boat slipped around the island and into a calm cove where we disembarked and picked our way toward the house that was set up on a low hill looking over the harbour. As the door opened, time rushed out. I felt the ghosts set free from their island purgatory as they pushed past us and out into the morning air. The building had not been a prison or a time capsule, but an ever-tightening loop, stacking its meanings in deep layers. Here, each object and nod to another time bristled with its own island energy. I almost felt everything call out to me, hoping for some kind of break from its loop. Every *thing* in this place had concentrated itself to become a miniature heterotopia (Foucault 1986), a world of islands within islands within islands.

From my perspective, it is important that we not consider ruins as sites of mourning and loss, at least not by necessity. To return to Deleuze and Guattari's (1988) concept of immanence, a ruin is more of a location of intersections and potential, a moment of moments in perpetual conversation. And while I do not discount the fact that ruins may have certain and specific emotional resonances for people, they are also productive sites of critical reflection and inquiry. What happens if we detach and reroute our embedded presuppositions of ruins in the same way that I have proposed that we reconsider our understanding of islands? Ruins can be islands, islands can be ruins, and both or neither of these views can be true.

In thinking *with* ruins in the context of islands, it is important to consider what surrounds them and what shorelines contain them. From my perspective, this force is time. A temporal ocean that separates ruin from the Mainlandish temporal flows and isolates ruins as a material Other that occupies a distinct temporal realm (Fabian 1983). If, in certain ways, islands are geographic and cultural isolates, then ruins can easily function as temporal places apart, severed from modernist notions of progress and set adrift to fade and die.

To my mind, both islands (at least in the sense of isolation) and ruins appear as castoffs of capitalism's voracious march toward mythologies of newness, efficiency, and progress. I think back to a walk I took with a Yapese woman through her village where she carefully pointed out the stone money, the dancing areas, and seating platforms used to host visitors from other villages, and the men's house, a central feature of all settlements on the island. Off in the jungle, a hundred metres or so from the main road, a seemingly abandoned structure quietly huddled into oblivion. I asked her what it had been used for. It had been a women's house, a place for menstruation rituals, conversations, and the production of everyday objects. I asked her if she felt sad about its demise. She looked at me somewhat quizzically and said that she was not upset, that it had simply fallen into disuse. And in this moment, I realized that the categories we ascribe and impose on our world are always-already divergent from local understandings. Ruins and islands are material facts, just as they are cultural constructs, and it is in this dynamic flow that we begin to see the value in *thinking with* these—literal and figurative—structures. Ruins, as do many other nodes in everyday life, offer another view to islands, an additional lens through which to see the accumulated layers, lines, and times.

## Material Worlds: The Half-Lives and Afterlives of Ordinary Things

My concept of object-oriented hauntology draws on recent work on *object-oriented ontology* (Harman 2018) and Derrida's (1994) notion of *hauntology* (the ideological archaeology that examines the power of the could-have-been future, an ontology of haunting) to create a practical ethnographic framework for understanding the human-object interface. Here, objects/artifacts/things become shorthand mediums for conveying nostalgia, longing, and the museumification of cultural forms. As discussed earlier, islands can be seen as an echo chamber for culture, its half-lives and afterlives reverberating back and forth like waves that forever pound the shorelines of their existence.

In light of the previous discussions of immanence (Deleuze and Guattari 1988), considering an object-oriented (and ethnographic) approach to hauntology might better be described as the pursuit of a location of convergence and dissolution, the move to a body without organs (Deleuze and Guattari 1988), where divisions and specificities mold themselves into a collective whole. And in some ways, this view starts to erase the heterotopic layers to form a pure Island wherein everything is always-already.

At this point I feel that the discussion needs some anchorage, a means of drawing it back to the lived realities of the island people, places, and things. So, while it is possible to reflect on these notions of bodies without organs and planes of immanence, it is also possible to apply these modes of thinking to questions of the essential and quotidian nature of islands. If we accept that there are certain ways of *thinking with* and *being on* islands—some anthropological, some that are otherwise—we can consider this framing in the following light: the work of the anthropologist of islands is to separate the stacked ontologies before then synthesizing them into a uniform whole that is formed and informed by the sum of its component pieces. Here, objects, places, and humans are at one point distinct, and at another point they are fused. What happens if and when we stop attempting to tease out our interactions with objects and see them as active agents in the world of humans? And islands are the perfect place to address this question because these divisions and interactions are bounded and deep, making for a much more immediate and resonant form of analysis.

# 9 A Conclusion by Means of Describing Certain Lessons That Islands Have Taught Me

## Lesson 1: Anchors

I have often found myself bound by the boundedness of islands, held in lines that offer situated reflections and insights that lose their blurry edges in time and place. Islands have allowed me to distil my awareness of the world.

Islands anchor me to my vocation. Being in these places centres and magnifies what it means to be an anthropologist of the Anthropocene. Here, I place myself in dialogue with my Self. The solitude calls for attunement and closer reading. Islands have taught me to be an anthropologist.

Islands anchor me to the landscape. As I've described in several other parts of this book, the layers of geography, ontology, and ideology that rise, fall, and intertwine form layers and lines that make a vertical world, a multiverse of islands that coexist. In essence, there is the lived and formed tactile island, the place of mosses and lava, of snow and sand, scraggly pine and swaying palm fronds. In all of my intangible meditations, I have learned to trust and value the fragility and power of islands, to remember their physicality and the ways that these landscapes make people, places, and things. I am reminded of Yi-Fu Tuan's (1974) notion of *topophilia* that describes the connection between human beings and the landscapes that they infuse with cultural significance. Islands have taught me to read these connections, and to recognize and define my own. The landscape becomes both text and context, and this affinity forms itself into human resonance and a deep attunement. Islands have taught me that ideas need earthly anchors.

Islands anchor me to writing. One of the primary challenges and goals of ethnography is the process of translating *there* into *here*, and the work of translating distant islands into relevant and relatable writing has made me grapple with the missing pieces, lost sentiments, and misinterpretations associated with travelling this line. The process of arresting a place, holding

DOI: 10.4324/9781003297581-9

it in suspension with words, is daunting and humbling, and it is forever incomplete (Clifford 2020). Islands have taught me how to begin to write.

## Lesson 2: Alone

In a reality of media oversaturation and cultural noise, of constantly splintering networks and fractal lifeworlds, islands can offer some respite from the waves of overabundance. Here, I do not claim that islands serve as an escape or retreat; for some the island is a prison, an inescapable inevitability. Still, I can only speak directly to my experience. It is important to note here that the island(er)s I have worked on/with reflect a certain realm of remoteness, often with relatively few inhabitants, and as such, offer greater opportunities for silence and solitude.

Islands have taught me to slow down, to quiet my emic and etic mind. Islands have given me the chance to remove some of the noise, to consider the phenomenology of being in the world. Every whispered note falls shining into my hand, and the wind clutches it, and I watch it fly away over the endless horizon of the North Atlantic or South Pacific. *Being* with islands has allowed me to *be* with myself, and to inhabit the echo chamber of existence a little more purposefully. Islands have taught me tactics for being alone (and being at home [Jackson 2000]) in the world.

## Lesson 3: Whirlpools

Travelling to, being on, and departing from islands has conditioned me to untie my expectations and ride the wave of uncertainty. Often, an island refuses to allow new plans or alternate routes. Often, the island has its own intentions for us.

Islands have taught me to improvise. And again, as the defined edges of these places create insightful loops of reflection, they also author whirlpools of chaos. This chaos was initially quite disarming, challenging me to release my grip on the answers to my preordained questions, to let go of what I thought I should find, and to experience the disruptions and reroutings as new texts and contexts, almost always richer and more dynamic.

Islands have given me a toolkit for wild acceptance. Islanders are masterful improvisers, something that I saw very explicitly during my many visits to Newfoundland's Fogo Island. Here, everything from language to furniture to foodways was the direct result of happenstance and bricolage (Levi-Strauss 1966). The islanders used what was directly at hand to forge a unique lifeworld. Embracing the swirled chaos of containment created some truly singular expressions of culture and identity. Islands have taught me to let the notes fall where they want to.

## Lesson 4: Wildernesses

There are more wildernesses than we can ever know. And like islands, there are ideological and geographic wildernesses. A wilderness is a place haunted by the absence of people. It is not always a place that has been without people, but one that is removed from them in the moment of encounter. The islands I have described in this book all contain their own wildernesses. The stones that flatten and stretch themselves across the faces of Iceland and Newfoundland, the careful rock pathways through Yap's jungled interior, and the barren face of the High Plains all form their own unpeopled dreamworlds.

Islands have taught me the importance of anthropology in the absence of people. Without other humans, the mythology of culture's counterpoint emerges in a new light. And as with everything I have previously outlined, wilderness on islands takes on a unique form, one that is often bound by a closed loop. An Icelandic wilderness takes its own shape that is wholly unlike that of a Yapese wilderness, and these specific qualities are the direct result of their islandness.

Before people, all islands were wilderness. At some point in the prehuman history of every island, everything was wilderness. Yet it is important, as Raymond Williams (1975) reminds us, to avoid the romantic idyllic imagination of island wilderness as some vanished Eden. It is simply a part of the half-life of places, an epoch like many more that will rise and fall. In a post-Anthropocene world, islands will return to their pre-human wilderness.

Islands have asked me to think with their wildernesses. They have asked me to consider humans in our absence and examine the depth of our presence, and to reflect on the category of wilderness as a cultural construction. Again, islands have given me an enclosure for compact immersion in the interplay between wild and domesticated places. Islands have taught me that culture is impermanent, and that the world is always more than human.

## Lesson 5: The Magic of Everyday Life

Seeing an island that I have previously visited through new eyes revives its magic and reminds me of how these places allow us to encounter and (often only briefly) inhabit parallel versions of everyday life. For the last several years, I have been taking groups of eight students to Iceland to participate in an intensive 2-week ethnographic field course. And while I have been travelling to Iceland regularly for over 15 years now, each time I see my students' anthropological wonderment well up and pour out, I am renewed in my affinity and attention for this place.

Islands have taught me about the magical relativity of everyday lives. And I suppose that this awareness is not limited to islands, but the wonderment they evoke seems somehow bound to their islandness. I'm not sure if I can apprehend the specific quality of this sentiment, but I am certain that seeing these places and the practice of anthropology through my students' eyes rekindles my sense of purpose and gives me hope for future anthropologists and writers. Yet why does this response bind itself to the context of islands? Why do I feel the resonance so much more palpably in these places? Why does the beauty of the everyday seem to crackle at the surface of everything in a way that on the mainland often seems deeply buried and less easily accessed? And once again I return to the line (Ingold 2007) and loop. The line has been drawn and has drawn me here, it draws itself around my students' imaginations and fascinations to form a feedback loop where layers of insight compound and accumulate. Each encounter with the island's people, places, and things adds new strata for instant reflection. As the everyday life of the island world delaminates itself inside its limits, new kinds of magic seep out and are carried back across oceans and miles.

Anthropology, for me, has always been about the beauty of the microtonal world of human culture, the careful attunement to the rhythms of the everyday caught in overlapping and intersecting loops. And it has also been about the privilege of being able to share these attentions and their associated tactics with my students. Islands have taught me to collect ordinary loops of culture to produce extraordinary viewpoints of everyday life.

# Afterword

## The Benefits of Thinking with Anthropology and Islands

My friend and fellow anthropologist Anne Brydon (2000) has described the practice of anthropology as "life skills for nomads", and for me this description embodies everything that I value in my continued pursuit of cultural meaning. Rather than the production of knowledge for its own sake, anthropology (at least when practiced in a conscientious and reflexive way) becomes a toolkit of assembled life skills, a collection of ways of being in the world.

My experiences as an anthropologist of islands have equipped me with a variety of tools and skills that serve me well in everyday life. I have become more patient, reflexive, adaptable, and grateful. For me, anthropology is more than an academic discipline, it is a framework for life. Sometimes it makes things more beautiful, sometimes more terrifying, but always more dynamic and richer. And we are certainly nomads in our constantly shifting attentions and inquiries, following the various means of meaning-making from site to site. Anthropology definitely provides us with the life skills we need to ask and address the critical questions of everyday existence in the Anthropocene. It offers us a glossary for beginning to understand the collected and entangled hows and whys of our humanness.

As sites for these dialogues, islands give us the chance to apply these life skills in a living laboratory. Be it a geographic island where loops are made from watery boundaries, or an ideological one bound by non-geographic forces, islands (and the ways that we imagine them) are places apart from the flows of the mainland everyday. And it is this quality that first drew anthropologists to conduct their research in these places; the island is the set and setting for a journey into the heart of the everyday. Along with this idea, I think it is important to note, as I have elsewhere in this book, that islands can be many things, and both islanders and mainlanders can build out these kinds of testing grounds. In some ways, anthropologists and anthropologically minded people are continually building and rebuilding their own private islands. They are carving them out of place, assembling them out of things, and populating them with people. So when we consider the island as something that is both very tactile and very ephemeral, we open up a world of infinite islands, an endless archipelago that disappears over an unreachable horizon. And here is where we can start to collect our nomadic life skills.

I truly hope that this book has offered a few new tools and life skills, for anthropologists, writers, and anyone improvising their way through the wild assembly of the Anthropocene.

# Bibliography

Adler, Renata. *Speedboat*. 1st ed. Random House, 1976.

Anderson, B.R.O. (2006) *Imagined communities: reflections on the origin and spread of nationalism*. Rev. ed. Verso.

Appadurai, Arjun. "Disjuncture and Difference in the Global Cultural Economy." *Theory, Culture & Society*, vol. 7, no. 2–3, 1990, pp. 295–310.

Augé, Marc. *Oblivion*. University of Minnesota Press, 2004.

Augé, Marc. *Non-Places: An Introduction to Supermodernity* (Translated by John Howe). 2nd ed. Verso, 2009.

Bakhtin, Mikhail M., and Michael Holquist. *Dialogic Imagination: Four Essays*. University of Texas Press, 1981.

Baudrillard, Jean. *Simulacra and Simulation*. University of Michigan Press, 1994.

Benjamin, Walter. *The Work of Art in the Age of Its Technological Reproducibility, and Other Writings on Media*. Harvard University Press, 2008.

Benjamin, Walter. *Illuminations: Essays and Reflections*. Houghton Mifflin Harcourt, 1968.

Berlant, Lauren Gail, and Kathleen Stewart. *The Hundreds*. Duke University Press, 2019.

Bohannan, Paul. "The Impact of Money on an African Subsistence Economy." *The Journal of Economic History*, vol. 19, no. 4, 1959, pp. 491–503.

Bourdieu, Pierre. "The Forms of Capital (1986)." *Cultural Theory: An Anthology*, vol. 1, 2011, pp. 81–93.

Brown, Bill. "Thing Theory." *Critical Inquiry*, vol. 28, no. 1, 2001, pp. 1–22.

Brydon, Anne. "Site Re-Visionings: On Seeing Space." *Space and Culture*, vol. 3, no. 4–5, Dec. 2000, pp. 187–203.

Certeau, Michel de. *The Practice of Everyday Life*. University of California Press, 1984.

Clifford, James. *Writing Culture*. Edited by James Clifford and George E. Marcus, University of California Press, 2020, pp. 1–26.

Deleuze, Gilles, and Félix Guattari. *A Thousand Plateaus: Capitalism and Schizophrenia*. Athlone Press, 1988.

Derrida, Jacques. *Specters of Marx: The State of the Debt, the Work of Mourning, and the New International*. Routledge, 1994.

DeSilvey, Caitlin. "Observed Decay: Telling Stories with Mutable Things." *Journal of Material Culture*, vol. 11, no. 3, 2006, pp. 318–338.

Edensor, Tim. "The Ghosts of Industrial Ruins: Ordering and Disordering Memory in Excessive Space." *Environment and Planning D: Society and Space*, vol. 23, no. 6, 2005, pp. 829–849.

Edmond, Rod, and Vanessa Smith, editors. *Islands in History and Representation*. Routledge, 2020.

Fabian, Johannes. *Time and the Other: How Anthropology Makes Its Object*. Columbia University Press, 1983.

Fitzpatrick, Scott M., and Jennifer Pinkowski. "Banking on Stone Money." *Archaeology*, vol. 57, no. 2, Mar. 2004, pp. 18–23.

Foucault, Michel. "Of Other Spaces." *Diacritics*, vol. 16, no. 1, 1986, pp. 22–27.

Friedman, Milton. *Money Mischief: Episodes in Monetary History*. Mariner Books, 1994.

Furness, William Henry, 1866–1920. *The Island of Stone Money: Uap of the Carolines*. J.B. Lippincott, 1910.

Geoffrey, Wall. "From Carrying Capacity to Overtourism: A Perspective Article." *Tourism Review*, vol. 75, no. 1, 2020, pp. 212–215.

Geertz, Clifford. Deep Hanging Out. *The New York Review of Books*. www.nybooks.com/articles/1998/10/22/deep-hanging-out/. Accessed 14 October 2021.

Gillis, John. *Islands of the Mind: How the Human Imagination Created the Atlantic World*. Palgrave Macmillan, 2009.

Glissant, Edouard. *Poetics of Relation* (Translated by Betsy Wing). University of Michigan Press, 1997.

Godfrey, Baldacchino. "Studying Islands: On Whose Terms? Some Epistemological and Methodological Challenges to the Pursuit of Island Studies." *Island Studies Journal*, vol. 3, 2008.

Gordillo, Gastón. *Rubble: The Afterlife of Destruction*. Duke University Press, 2014.

Harman, Graham. *Object-Oriented Ontology: A New Theory of Everything*. Pelican, Penguin Random House, 2018.

Hastrup, Kirsten. *A Place Apart: An Anthropological Study of the Icelandic World*. Clarendon Press, 1998.

Hau'ofa, Epeli. "Our Sea of Islands." *The Contemporary Pacific*, vol. 6, no. 1, 1994, pp. 148–161.

Hayward, Philip. "Aquapelagos and Aquapelagic Assemblages." *Shima*, vol. 6, no. 1, 2012, pp. 1–11.

Ingold, Tim. *Lines: A Brief History*. Taylor & Francis Group, 2007.

Jackson, Michael. *At Home in the World*. Duke University Press, 2000.

Kawabata, Yasunari. *Palm-of-the-Hand Stories*, 1988.

Kincaid, Jamaica. *A Small Place*. 1st ed. Penguin, 1989.

Korzybski, Alfred. *Science and Sanity: An Introduction to Non-Aristotelian Systems and General Semantics*. International Non-Aristotelian Library Publishing Company, 1933.

Kristín, Loftsdóttir. *Crisis and Coloniality at Europe's Margins: Creating Exotic Iceland*. Routledge, 2019.

Lessa, William A. "Ulithi and the Outer Native World." *American Anthropologist*, vol. 52, no. 1, 1950, pp. 27–52.

Levi-Strauss, Claude. *Structural Anthropology*. Rev ed. Basic Books, 1974.

Levi-Strauss, Claude. *The Raw and the Cooked* (Translated by John Weightman and Doreen Weightman). University of Chicago Press, 1983.

Levi-Strauss, Claude. *The Savage Mind*. University of Chicago Press, 1966.

Liesemer, Dirk. *Phantom Islands: In Search of Mythical Lands* (Translated by Peter Lewis). Haus Publishing, 2019.

Lury, Celia. "The Objects of Travel." *Touring Cultures*. Routledge, 2002, pp. 85–105.

MacArthur, Robert H., and Edward O. Wilson. *The Theory of Island Biogeography*. Rep ed. Princeton University Press, 2001.

Malinowski, Bronislaw. *A Diary in the Strict Sense of the Term*. Stanford University Press, 1989.

Malinowski, Bronislaw. *Argonauts of the Western Pacific: An Account of Native Enterprise and Adventure in the Archipelagoes of Melanesian New Guinea*. G. Routledge & Sons, E.P. Dutton & Co., 1922.

Malinowski, Bronislaw. *Magic, Science and Religion and Other Essays*. Waveland Pr Inc, 1992.

Malinowski, Bronislaw. *Sex and Repression in Savage Society*. 2nd ed. Routledge, 2003.

Mayda, Chris. "Resettlement in Newfoundland: Again." *American Geographical Society's Focus on Geography*, vol. 48, no. 1, 2004, pp. 27–32.

McLuhan, M. and Powers, B.R. (1992) *The global village: transformations in world life and media in the 21st century*. Oxford University Press.

Nemeh, Katherine H., and Jacqueline L. Longe, editors. "Ecotone." *The Gale Encyclopedia of Science*, 6th ed., vol. 3, Gale, 2021, pp. 1492–93.

Oliveros, Pauline. *Deep Listening: A Composer's Sound Practice*. iUniverse, Inc., 2005.

Ortner, Sherry. "On Key Symbols." *American Anthropologist*, vol. 75, no. 5, 1973, pp. 1338–1346.

Ortner, Sherry B. *Anthropology and Social Theory: Culture, Power, and the Acting Subject*. Duke University Press, 2006.

Oslund, Karen, and William Cronon. *Iceland Imagined: Nature, Culture, and Storytelling in the North Atlantic*. University of Washington Press, 2011.

Pálsson, Gísli. "Situating Nature: Ruins of Modernity as Náttúruperlur." *Tourist Studies*, vol. 13, no. 2, 2013, pp. 172–188.

Radcliffe-Brown, A. R. *The Andaman Islanders*. Cambridge University Press, 2013.

Rahimi, Sadeq. *The Hauntology of Everyday Life*. Palgrave Pivot, 2021.

Roni, Horn. *Island Zombie*. Princeton University Press, 2020.

Ronström, Owe. "Finding Their Place: Islands as Locus and Focus." *Cultural Geographies*, vol. 20, no. 2, 2013, pp. 153–165.

Sæþórsdóttir, A., Hall, C., and Wendt, M. "Overtourism in Iceland: Fantasy or Reality?" *Sustainability*, vol. 12, 2020, p. 7375.

Sahlins, Marshall. *Stone Age Economics*. 1st paperback ed. Aldine, 1974.

Schalansky, Judith. *Atlas of Remote Islands: Fifty Islands I Have Never Set Foot On and Never Will*. Translated by Christine Lo, First Edition, Penguin Books, 2010.

Sebald, W. G., *The Rings of Saturn*. New Directions, 1999.

Shelton, Allen. *Dreamworlds of Alabama*. University of Minnesota Press, 2007.

Stephen B. M., and Scott M. Fitzpatrick. "Banking on Stone Money: Ancient Antecedents to Bitcoin." *Economic Anthropology*, vol. 7, 2019, pp. 7–21.

Stewart, Kathleen. *Ordinary Affects*. Duke University Press, 2007.

Taussig, Michael. *Palma Africana*. University of Chicago Press, 2018.

Tsing, Anna Lowenhaupt. *Friction: An Ethnography of Global Connection*. Illustrated ed. Princeton University Press, 2005.

Tuan, Yi-fu. *Topophilia: A Study of Environmental Perception, Attitudes, and Values*. Prentice-Hall, 1974.

Walter, Benjamin. *One-Way Street: And Other Writings*. Verso, 2021.

Williams, Raymond. *The Country and the City*. Oxford University Press, 1975.

# Index

abandonment 30, 48, 54–55
anthropology x–xi, 2, 8–9,
    17, 24, 30, 59, 61–62;
    authentic subject of 42–43;
    history of 4–6
aquapelago 14, 17–18, 25

ecotones 9–10, 15, 19
ethnography ix, 8, 16,
    21–22

Faroe Islands 20–25
ferries 10–11

ghost town 45–49, 54–55
globalization 9–10, 42, 48

hauntology 50–56
heterotopia 3, 43, 50, 52, 54

Iceland 33–44; Westfjords region
    34–36, 38–39

islands 1–11, 61–62; and climate
    change 51; and material culture
    55–56; and narrative 6

lines 22–23, 27, 46, 60

micro-essays ix–x

Newfoundland 26–32

phenomenology 30–32, 52, 58
positionality xii, 16, 24, 58

resettlement 28

supernatural 16, 52–53

tourism 40–44

wilderness 59

Yap 13–18